DREAMS, NEUROSCIENCE, AND PSYCHOANALYSIS

Dreams, Neuroscience, and Psychoanalysis sets out to give a scientific consistency to the question of time and find out how time determines brain functioning. Neurological investigations into dreams and sleep since the mid-twentieth century have challenged our scientific conception of living beings. On this basis, Kéramat Movallali reviews the foundations of modern neurophysiology in the light of other trends in this field that have been neglected by the cognitive sciences, trends that seem to be increasingly confirmed by recent research.

The author begins by giving a historical view of fundamental questions such as the nature of the living being according to discoveries in ethology as well as in other research, especially that which is based on the theory of the reflex. It becomes clear in the process that these findings are consistent with the question of time as it has been considered in some major contemporary philosophies. This is then extended to the domain of dreams and sleep, as phenomena that are said to be elucidated by the question of time.

The question is then raised: can dreaming be considered as a drive? Based on the Freudian discovery of the unconscious and Lacan's teachings, Movallali seeks to provide a better understanding of the drives in general and dreams in particular. He explores neuroscience in terms of its development as well as its discoveries in the function of dreaming as an altered mode of consciousness. The challenge of confronting psychoanalysis with neuroscience forces us to go beyond their division and opposition. Psychoanalysis cannot overlook what has now become a worldwide scientific approach. Neuroscience, just like the cognitive sciences, will be further advanced by acknowledging the desiring dimension of humanity, which is at the very heart of its being as essentially related to the question of time. It is precisely this dimension that is at the core of psychoanalytic practice.

Dreams, Neuroscience, and Psychoanalysis will appeal to psychoanalysts and psychoanalytic psychotherapists as well as neuroscientists, psychologists, ethologists, philosophers and advanced students studying across these fields.

Kéramat Movallali is a psychoanalyst treating adults in private practice in Paris, France. He also conducts psychoanalytic treatments for children and adolescents in public institutions.

DREAMS, NEUROSCIENCE, AND PSYCHOANALYSIS

Mind, Body, and the Question of Time

Kéramat Movallali

TRANSLATED BY ANDREW WELLER

LONDON AND NEW YORK

First published 2017
by Routledge
2 Park Square, Milton Park, Abingdon, Oxon OX14 4RN

and by Routledge
711 Third Avenue, New York, NY 10017

Routledge is an imprint of the Taylor & Francis Group, an informa business

© 2017 Kéramat Movallali

This book is a translation of a work previously published in French as
*Contribution à la Clinique du Rêve: la Psychanalyse et les Neurosciences
(Psychanalyse et Civilisation)*, L'Harmattan (2007).

Translation into English by Andrew Weller

British Library Cataloguing in Publication Data
A catalogue record for this book is available from the British Library

Library of Congress Cataloging in Publication Data
Names: Movallali, Kéramat, 1948- author.
Title: Dreams, neuroscience, and psychoanalysis : mind, body, and the
    question of time / Kéramat Movallali. Other titles: Contribution áa la
    clinique du rãeve. English
Description: Abingdon, Oxon ; New York, NY : Routledge, 2017. |
    "This book is a translation of a work previously published in French
    as Contribution áa la Clinique du Rãeve: la Psychanalyse et les
    Neurosciences (Psychanalyse et Civilisation), L'Harmattan (2007)." |
    Includes bibliographical references and index.
Identifiers: LCCN 2016036347| ISBN 9781138858244
    (hardback : alk. paper) | ISBN 9781138858251 (pbk. : alk. paper) |
    ISBN 9781315718156 (e-book : alk. paper)
Subjects: LCSH: Dream interpretation. | Dreams—Physiological aspects.
Classification: LCC BF175.5.D74 M6913 2017 | DDC 154.6/3—dc23
LC record available at https://lccn.loc.gov/2016036347

ISBN: 978-1-138-85824-4 (hbk)
ISBN: 978-1-138-85825-1 (pbk)
ISBN: 978-1-315-71815-6 (ebk)

Typeset in Times New Roman
by Swales & Willis Ltd, Exeter, Devon, UK

CONTENTS

ILLUSTRATIONS

PREFACE

The scientific approach to the neurophysiology of dreams only dates back to the middle of the twentieth century. It was then that Nathaniel Kleitman and Eugene Aserinsky succeeded in establishing the connection between dreams and rapid eye movements during sleep. To be exact, this discovery did not concern dreams as such, but their neurological mechanisms. Scientists were then led to dismiss the idea according to which sleep occurs when "the flashing shuttles" of the brain begin gradually to go out so that the mind, which is supposed to be like an "enchanted loom", can reach an optimal point of rest (Sherrington, 1942, p. 178). During the same epoch, Freud upheld the contrary view, regarding sleep as the high place of mental activity.

Following the historic discovery by Kleitman and his pupil, neurophysiological laboratories throughout the world began to study the relation between eye movements and dreaming in greater depth. From that point on, the Freudian theory of dreams became the privileged target of attacks and refutations. Wish fulfilment and the unconscious mechanisms that govern it had no equivalence or counterpart in these electroencephalographic research studies. It was nonetheless the psychoanalytic theory of *dreams* that was targeted and not Freud's original approach concerning the dream process itself which, for its part, was in many respects in keeping with the new neurophysiological research. Thanks to the ambiguity of the term "dreaming", this research tended to confuse dreams and sleep, whereas Freud had clearly established the connection between them. According to him, the dream makes use of the altered state of consciousness during sleep in order to have access to unconscious desire. This is one of its principal functions. Researchers in the neurophysiology of sleep, such as Robert McCarley and Allan Hobson, who manifested their open hostility to Freud's ideas, maintained the contrary. According to these authors, the dream is a superfluous neurological production without importance, a sort of roaming of the nervous cells at the level of the brainstem. The forebrain more or less takes charge of these cells and tries to give them some meaning by drawing on the dream's day residues. Other researchers, constituting a minority, sought to prove the contrary by assigning the forebrain with the principal role in the occurrence of dreams without, however, interesting themselves in psychoanalytic theory. It was necessary to wait for the contributions of other researchers to realize that one could not simply assign the immense mental activity required in dream formation to one part of the brain.

It is probably too soon to judge, but one would not be mistaken in asserting that *the study of dreams is likely to turn the whole field of investigations in the brain sciences upside down.* Contrary to the postulate of neuroscience according to which the encephalon is constituted at birth as a mere receptacle of the external world, research contributions into the neurophysiology of dreams have constantly shown the contrary, discarding the dominant theory based on the *tabula rasa* of the mind. We shall see throughout this book how scientists came, not without surprise, to recognize that the brain was first and foremost endowed with an intrinsic and self-referential system. What was even more astonishing for them was that the pre-existent constituents of the brain are far from deprived of contact with the external world, since they have been acquired during the course of evolution from this same ambient world, not by means of simulation as is generally believed, but by actively emulating "the living" (*le vivant*) in its intimate interaction with external stimuli (*Umwelt*). Ethology has really taught us a great deal on the subject of the living. That is why I will be giving it a prominent place in the first part of this book. The question of the living and of its fundamental correlate, time, is approached in it from the angle of philosophy but also from that of the life sciences. It is in this connection that I will be led to speak of a renewed trend in neurophysiology that places temporality at the heart of every aspect of the functioning of the organism.

We shall also see how the most recent research, drawing on the neurophysiology of sleep, has succeeded in contributing decisive elements of response to the question of consciousness. Just like Freud, researchers have come to consider the dream as a curious alteration of consciousness. The exceptional state of attention and memory emphasized in these investigations tends to give credit to the idea of an increased activity of unconscious impulses during paradoxical sleep when the majority of our dreams occur. And so, it would not be surprising if one day soon, by virtue of these discoveries, we can approach the Freudian unconscious on different bases or along other lines that are unsuspected today.

A recent current of research in neurobiology has in fact developed out of an original approach that has remained a minority position since the beginning of the twentieth century. Drawing on scientific experiments, it puts temporality at the centre of the inquiry concerning the physiological functioning of the living. It established itself by rejecting from the outset the Pavlovian principle according to which the nervous system is based on the model of the reflexes and their conditioning.

We know that the neurophysiology of the twentieth century was, and still continues to be, inspired by the reflex arc. Behaviourism made a great deal of it by dismissing any means of recourse apart from the stimuli/response system for studying the living. Notwithstanding its refutation of the behaviourist approach, cognitivism remained a prisoner of it by denying, in its turn, any intrinsic constitution to the human soul. Its quest for objectivity on the model of the exact sciences removes its possibility of turning towards a possible subjectivity in living beings in general and in man in particular. It is always behaviour that remains its sole means of investigation. Cognitive psychology is thus condemned to conduct research that denies all interiority to the individual from the moment he is said to be more than the sum of his visible and calculable behaviour. This rejection of interiority is translated, on the level of social and individual life, by the principle of conformity, whereby any original behaviour is likely to be considered deviant in relation to the established norms.

Let us return now to the recent current of research, which is indeed a minority one, in neurophysiology which is distinguished by its opposition to Pavlovism. It reveals in a new light the remarkable dynamism of a living being, which, far from being passively at the mercy of the external world, shapes it in its own manner while forging itself in its image. It intervenes in it as much as it depends on it. This requires, and we will see how, the living to be endowed with a temporal aptitude which does not exist as such in the purely physical universe. This approach to temporality led the advocates of this recent current to interest themselves in the highly elaborate reflections of the phenomenon of time in the work of philosophers of phenomenological inspiration. I will attempt to show, in the course of this book, the singular proximity between the phenomenological approach and that which prevails in this branch of research. We will also find that the theme of temporality is part of their specific approach.

I shall therefore be taking temporality as a principal purpose of the dream process. By inquiring into neurophysiological research, the discoveries in ethnology, as well as philosophical investigations, I will try to draw on the clinical psychoanalytic experience of dreams in order to locate time at the centre of any study concerning the human soul. This will lead me to put forward a new theory of dream activity, based on the postulate of a specific temporality which is in keeping both with the Freudian discovery and with Lacanian articulations. As for the neurophysiologist, we will see that far from being in contradiction with psychoanalytic clinical practice, his investigations may be said, in fact, to back it up.

The enormous expansion of the sciences since the Renaissance results from the fact that our gaze has turned on all sides towards external reality. Will the central place that is occupied today by research into dream activity succeed in decentring our investigations in such a way as to question this reality from the perspective of our dark nights, which seem to give us access to it in a different way? It is precisely such a challenge that was the inspiration behind this book.

I have endeavoured to write this book in a manner that is accessible to a wide public. I would like lay readers to be able to find their mark in it just as much as the specialist, merely as dreamers who may find themselves challenged by my questionings throughout the different peripeteia of the book. In order to facilitate their task, I have included nearly one hundred clinical examples and dreams derived from my own clinical practice each time that it seemed necessary.

Fontenay Saint Père
Summer 2014

PART I
Body/mind

1

WHAT IS THE LIVING?

The sciences say little about what distinguishes man from animals. Is this a historical error or a unique aptitude in the evolution of science to assimilate or identify the functioning of the human brain, albeit assigning a greater complexity to it, with that of animals? Such questions oblige us to turn towards the conception that we have of man today.

It is as a *living* being that man is apprehended by neuroscience. We owe this conception to the modern age which discarded the idea of man constituted by metaphysical thought. The question concerning the essence of life has nonetheless been left in the shadows. This is neither a matter of negligence nor of forgetting. Science sets out to study the mechanisms and the modes of functioning of living matter; it cannot answer a question that remains outside its purview. But this does not mean that science does not possess its own conception of it. Is *life,* insofar as it constitutes the entity called living, a modern conception that has come into the world since the advent of biology? Let us retrace briefly what may have paved the way for such an event.

"By life", Aristotle (1952) writes in *De Anima* (*On the Soul,* II, 1), "we mean self-nutrition and growth (with its correlative decay)." In Aristotle, life tends to be identified with the soul. This gives rise to the etymological redundancy that is the *animal soul*, which separates the animated body from inanimate matter.

Aristotle establishes a difference between animal life (ζôé), enjoyed by both the gods and men, (as well as animals) and the *mode* of human life (βιος), which encompasses the manner in which we accomplish our lives as common mortals. With the advent of science, the βιος was assimilated with life in the inclusive sense of the entity that refers to the ensemble of living beings. And so it acquired surreptitiously an eminently political meaning as an entity that could henceforth be treated *en masse* (see Foucault, 2004; and also Agamben, 1997).

The doctrine of the "animal-machine" (Descartes) marked the beginning of a radical change in the seventeenth century. Around the same time, illness and health became the principal *criteria* concerning life, which then constituted an almost exclusively medical issue. It was Georg Ernst Stahl (1708), a Prussian doctor, who would argue that there was an intrinsic relationship between *life and medicine* (see Canguilhem, 2002, pp. 15–31). According to him the soul is the vital principle that resists corruption, decomposition, and death. Leaving the idea of the soul out of account, Bichat (1994), in his *Recherches physiologiques sur la vie et la mort*, defined the conflict that opposes life and death.

We are still a long way from Freudian thought according to which the life drive tends intrinsically towards the death drive. Lamarck was the first to use the term *biology,* around the year 1800, in his book on hydrogeology (see Corsi, 2001). According to him, studies concerning living bodies should turn their attention to the simplest of organisms. This announced the immense progress that the biosciences would accomplish in the twentieth century in the domain of microscopic beings. This cellular conception would henceforth govern research contributions in the biological domain. Lamarck held that heat, the first act of life, determined the material soul of living bodies. With Cuvier we witnessed the birth of the *constancy principle* to which Freud was to refer in his drive theory. Life and death stand in a complementary relationship in the service of the principle of constancy. The latter guarantees, according to Cuvier, the incessant metamorphosis to which living bodies are subjected. Cuvier's *vitalism* introduced in a decisive manner the notion of vital forces. His conception of life was defended still more rigorously by Claude Bernard (1865) in his *Introduction to the Study of Experimental Medicine.*

The mechanistic theory is at the opposite pole to vitalism. It begins with the Cartesian idea that conceives of the animal as an automat. In the sciences it would prevail over vitalism. The latter made a radical distinction between vital processes and physico-chemical phenomena. Claiming allegiance to Aristotle, and above all to Hippocrates, the vitalist adepts of the Montpellier school in the eighteenth century, with Josef Barthez and Théophile de Bordeu as leading figures, resisted seeing their investigations being simply annexed by the sciences of inanimate matter. Vitalism was founded on the idea of an indeterminate force radically different from physico-chemical processes. For Xavier Bichat, vital properties, in respect of their variability and their intensity, were fundamentally opposed to inanimate nature which possessed invariable and fixed properties. Admittedly, Bichat rejected the principle of a single force directing all the functions of life; nonetheless, he thought that at the heart of nature, governed by physico-chemical laws, life represented a permanent exception (see Huneman, 1998). To this specificity of the living, the mechanistic theory would oppose the *law of inertia*, which determines the inanimate world and living beings alike. Vitalism was the last breath of a thousand-year-old way of thinking that was still resistant to the scientificity of the modern age. It attempted in vain to maintain the belief in the mysterious forces of living nature, even if it was ready to assign them verifiable scientific properties.

Underlying the law of inertia, which is so essential to the mechanistic conception, is the Cartesian *cogito* (the "I think, therefore I am") announcing the birth of the *subject of science*. This is why Lacan (1966a) traces the discovery of the Freudian unconscious back to Descartes' thought. In effect, the *cogito* is the historical moment when the gap between *knowledge* and *truth* was established. Science was henceforth founded on the quest for certainty without a concern for truth, relegating the latter to the register of faith. By introducing the law of inertia at the heart of his conception of the drives, Freud distanced himself from the outset from any possible vitalist implications concerning his nascent theory. The dualism of the life and death drives was also intended to counter the vitalists' belief in a single and unique vital force determining the living world. According to Robert McCarley, an American neurophysiologist, it was probably in reaction to such a belief that Freud immediately avoided endowing the psychical apparatus with endogenous properties. "Perhaps it was to avoid 'vitalistic' notions that Freud, in the *Project*, conceptualized neurons as passive transmitters and reservoirs of energy derived from outside the brain" (McCarley, 1998, p. 118). We know, more generally, that at the time of the *Project* (before 1900), Freud relied on existing neurological research. The latter claimed allegiance to the theory of conditional reflexes which had become the corner stone of the studies undertaken by Pavlov (in Russia) and Sherrington (in England). This theory, excluding any form of endogenous property in the brain, was demolished by the new neurology from the 1950s onwards. I will come back to this later when studying anticipative neurophysiology.

Cybernetics is the most accomplished outcome of the mechanistic theory. During the course of the twentieth century, cognitivism, and especially cybernetics, both claiming to have invented artificial intelligence (the computer) on the model of the human brain, would provide the conceptual tools necessary for neuroscience which, in turn, would inspire the connexionists to conceive of the "formal neurons" of computers. We will also return to this later.

As I have said, it was partly against vitalist ideas that Freud conceived of the death drive as being intrinsically linked to the life drive. The phenomenon of *repetition*, which is inherent to it, constitutes the very foundations of clinical psychoanalysis. The latter is based on the *transference*, that is, the repetition of the patient's past on to the analyst's person.

However, the death drive has not as yet found a consequent echo in scientific research. Going beyond the biological framework, it would be more likely to contribute to the understanding of the civilizing dimension of man. As evidence of this one may cite the late works of Freud such as *Civilization and Its Discontents* (1930). And yet every clinician knows from experience the extent to which the notion of the death drive is capable of explaining extremely serious phenomena that could scarcely be apprehended otherwise. Rupture, conflicts inherent to the system, unforeseen events, disunion or fragmentation, disaster and irruption are all manifestations governed by this deadly drive.

Cybernetics or information theory has tried to impose another conception of living beings. I am speaking of a science of systems control equipped with entries/exits and governed by the principle of maintaining a constant balance between the incoming and outgoing elements. It was created in 1948 by the American mathematician Norbert Wiener on the model of thermodynamic apparatuses. His concepts of reciprocal action and feedback rest on a system that detects gaps with a view to regulating and integrating them within an organization whose functioning depends on the principle of homeostasis. The system is thus dependent on a *signal* that refers to an ensemble of pre-established connections and communications in order to conserve the organization in its totality. The signal in question tries to remedy the disorder and the imbalance that has occurred in the system so as to reestablish the organizational order of the living, conceived of as a cybernetic machine. Cybernetics announced the beginning of a conception treating both living beings and machines as self-organizational systems. It contributed eventually to the construction of artificial intelligence (computers).

Freud, too, drew inspiration in more ways than one from such models. Concepts such as signal anxiety or *entropy* as an effect of repetition served as models to explain a certain number of phenomena at the heart of psychoanalytic theory.

Birth of a planetary science

Cognitivism is linked to a major signifier: an intelligent machine or programmable computer. It grew out of *mathematical formal logic*, which was in vogue in the years 1920–1940. But the origin of this formal logic goes back to the work of the philosopher and mathematician Gottlob Frege (1848–1925) who, with the invention of his ideogram (*Begriffsschrift*) hoped to dispel what he saw as the confusion of natural languages. It was with the help of this ideogram that he tried, in the wake of Leibniz, to promote a universal language capable of determining, by means of radical logic, the falseness of propositions (see Frege, 1971). His logicist system was aimed at formalizing thought. It was in this respect that he influenced the entire computationist current of the cognitive sciences in the twentieth century, based on the model of his ideogram which has become the artificial intelligence of computers. Frege's logicism was integrated within the *logical empiricism* of a certain number of positivist philosophers such as Moritz Schlick, Ludwig Wittgenstein, Bertrand Russell, George Edward Moore, Henri Poincaré, and Kurt Gödel, who all belonged to the famous *Vienna Circle.* Hence their essential idea consisted in only attributing the status of scientific truth to that which is either formal or logical (see Ouelbani, 2006).

Enriched by the immense research studies carried out during these years in the domain of formal logic, cybernetics began to attract all those who thought that by simulating the intellectual aptitudes of the human being it might finally be possible to build Descartes' animal-machine. But before this could be achieved it was necessary for researchers to distance themselves from the mechanistic aims of the philosopher and, while remaining in the materialist domain, make use of the high degree of generality that computer science procured for the machine.

It was in the 1940s that Alan Turing (1912–1954), an English logician and engineer, laid out the conceptual basis of the cognitive sciences. According to him, the latter should not only be able to describe and explain the aptitudes of the human mind, but also to *simulate* them through the agency of automata created for this purpose. After playing a pivotal role in discovering the codes governing the military communications of the Nazi army in favour of his country, Alan Turing was deprived of his civil rights on account of his homosexuality. The apple filled with cyanide, with which he committed suicide, represented this unforgivable sin in the eyes of the British authorities. It was to become the emblem of one of the most prestigious computer companies.

Cybernetic theory was no longer restricted to its essential elements, namely, homeostasis, feedback, and the "dementalized" contents of psychic phenomena. It now made use of mathematical and computing logic, which allowed it to establish the equation between the machine and the mind. The death knell of behaviourism, which had conceived of the brain as the "black box" that was of no concern to the scientist, was sounded. The stimulus/response system proved insufficient, for it was now possible to focus on the "black box" with a view to studying its formal functioning. Cognitive science was thus born, equipped with a major tool, the computer, which was supposed to be able to *simulate* mental processes in the formalist sense of the term.

Linguistics, neurophysiology, and experimental psychology played their part in promoting an interdisciplinary theoretical corpus during the "Macy" Conferences, held between 1945 and 1953, with the presence of eminent researchers such as Wiener, John von Neumann, Rosenblueth and Walter Pitts. To this we may add the undeniable contribution of McCulloch of the *Massachussetts Institute of Technology* (M.I.T.). From the 1960s onwards cognitivist theory went from strength to strength. The majority of the behavioural psychologists converted to it gradually, but we know that behaviourism is far from having been defeated. Its spirit continues to determine cognitive psychology, as can be seen from cognitive-behavioural therapies (CBT), which sometimes leave clinicians powerless in the face of the havoc they can cause as a therapeutic tool.

Cognitive-behavioural therapies

As they are techniques that are essentially focused on the symptom, CBT do not consider the psychic organization of the individual as a whole. It is for this reason that we often witness the displacement of the symptom treated, giving rise to other psychic problems that may manifest themselves surreptitiously in the medium or long term. Given that the displacement of the symptom may occur without the patient's knowledge and in an unrecognized form, studies carried out to verify the reliability of these techniques do not always constitute irrefutable proof. These studies of verification are to be treated with all the more caution in that they set out to compare different types of therapy, even though each of them pursues different objectives. By focusing on symptoms to make such comparisons, the studies in question cannot be justified in psychic matters where each symptom possesses its own subjectivity and its different underlying causes. Equating psychic symptoms with nosographical entities is clearly an ideological error.

The report by INSERM (2004) concerning the reliability of the different psycho-therapeutic methods may be said to be based, precisely, on such an error, for it relies on criteria on the basis of which the DSM-IV reduces the symptom to a *disorder*, an eminently controversial nosographic (and not nosographical) concept.

Cognitive-behavioural techniques pay little attention either to the psychic structure of patients or to their history. For cognitivists, their suffering is only a symptom involving their behaviour rather than their person. And so the belief in the behaviourist's *black box* continues to persist in a different guise. What cognitivists are interested in is the observable and measurable behaviours of individuals and not their provenance. The patient's psychic organization is considered as an agglomerate of disorders conceived of as a deviation from the established norms with which everyone is supposed to conform. Like a certain form of medicine from which they draw inspiration, cognitive therapies are more focused on the symptom than on the patient. The question of the transference, that is, of the manner in which the patient cathects the person of the therapist, is ignored, even though it determines how the therapy unfolds. Likewise, the cognitivist clinician refuses to dwell on his (or her) own countertransference towards the patient. He hides behind his therapeutic technique, overlooking himself as a person and overlooking his patient as a whole subject. This negligence cannot fail to compromise, in the medium or long term, the therapeutic result. The question of the patient/therapist relationship is completely blocked out in order to preserve the "scientificity" of the technique and to avoid its non-measurable character in the clinical evaluation.

Cognitive therapies require scarcely any empathy on the clinician's part. In the absence of any work of supervision on the countertransference of the person of the therapist towards his patient, empathy boils down either to a superficial relationship of courtesy or to a phobic reaction to entering the patient's private space. The risk of the therapist getting inextricably bogged down in the relationship with the patient is by no means negligible. Without having recourse to a diagnosis based on clinical experience that takes into consideration the patient's psychic structure, the therapist is at best guided in his approach prior to the therapy proper by the disorders listed in the *Diagnostic and Statistical Manual of Mental Disorders* (DSM).

As its name indicates, the DSM is established on the basis of statistics collected on patients. These are put together from findings furnished jointly by hospitals and the army in the United States. The listed disorders correspond, with little difference, to the *International Classification of Diseases* (ICD) published by the World Health Organization (WHO).

The fragmentation of nosological entities into isolated disorders could only serve to encourage pharmaceutical companies, which soon took advantage of it. "We know today," Steeves Demazeux (2013) writes, "how strong the collusion is between the decisions of the DSM and the interests of the pharmaceutical industry. Nearly 70% of the experts who work for DSM-5 have had financial links during their recent career with the pharmaceutical industry" (p. 19).

This kind of conflict of interests is an edifying example of what Michel Foucault (2004) once called biopolitics.

Another more frequent means for the cognitivist therapist to establish his diagnosis is to refer to the *Global Assessment of Functioning* (GAF), which is nothing other than the Axis V of the fourth edition of the DSM. It is a numerical descending scale from 0–100 going from mental health (91–100) to mental illnesses (1–10). Everything is rated and measured to evaluate the patient with a view to making a diagnosis. No supervision or personal psychological work on the part of the clinician is required, thus sparing him the need to make any subjective judgement.

Cognitive therapies are essentially based on a simple technique which consists, since Pavlov, in deconditioning the patient's symptom, on the condition that it does not mobilize the patient's unconscious forces. But there are numerous such cases. I need do no more here than to report the case of a young woman suffering from problems of body image. During a session of group supervision, I had given a clear warning to the cognitivist psychologist who had begun a process of deconditioning and explanation with the young woman with the aid of a mirror and video recordings. The description that the psychologist gave of her was more suggestive of a fragile structure linked to underlying psychotic difficulties. A short while after, the patient's family was obliged to intern her in a psychiatric clinic. The techniques employed by the psychologist had probably touched on an organization whose current symptom was far from being simply a matter of a problem of body schema.

It must be admitted, however, that CBT are efficient when it comes to symptoms that are related to what may be called primary defence mechanisms. These are symptoms that may or may not be related to traumas and which affect individuals in a moment of intense fragility. If it is unable to mobilize sufficient psychic resources, the mind will employ massive defensive means such as phobias and panic states. Both these nosographical entities are characterized by their lack of mentalization. They evoke first and foremost the neurological mechanism which, as it were, reacts blindly, engendering terror and disarray in a mode that is devoid of mental elaboration.

CBT respond "marvellously" to these kinds of disorders which, because words and speech are lacking, cannot easily be linked up with the subjective history of the patient. Clinically, what is visible, primarily, is a psychical apparatus whose essential task is to deal with the ongoing factual affairs of the external world. This particular mode of functioning is now referred to as "operational thinking" (*pensée opératoire*; see Marty, 1998), where the banality and platitude of saying (Sami-Ali, 1980) seals every inner event under a lead cloak. The patient's communications will be limited to an account of the facts whose poverty of description offends the clinician's sensibility. In the absence of an inner psychic life, one gets the impression, from what the patient says, that everything is a result of external causes. Moreover, action and acting out constitute the patient's favoured mode of functioning. Consequently, what we witness is the symbolic deficiency of the patient, namely his (or her) difficulty in integrating his stress and emotive shocks within a network of signifiers so that his subjective experience can be mobilized by appealing to his memories and his history. Faced with the uncertainties of the individual's life, this symbolic indigence may have effects both upstream and downstream. In both cases, one may find oneself faced with a particular neurological mechanism called "kindling" (Post, 1992, p. 1004). It consists of the recurrence of symptoms of mood disorders (depression, panic attacks, anxiety, phobias . . .) occurring at ever shorter intervals before finally becoming a neurological automatism, often resistant to all forms of treatment. Over and beyond any clinical arguments, it is no doubt towards anthropology that we would have to turn to see the effects of the digital era and the advent of Descartes' animal-machine on the massive resurgence of the automatism in question (Neidich, 2013). Indeed, the latter characterizes a considerable portion of current psychopathological affections. Without such investigations at the level of the historicity of the forms and structures that these pathologies assume, we risk getting bogged down in interminable quarrels and controversies between the supporters of diverging schools of thought. It goes without saying that it is from the same anthropological perspective that we need to question the function and the place of "operational thinking" nowadays.

Highlighted in psychosomatic symptoms in the course of the 1960s (see Marty and M'Uzan, 1963) operational thinking seems to have gained ground to the point of invading the clinical space of the psychologist today. It is not surprising, therefore, that we are witnessing a massive recourse to behaviourist techniques. These do not require the patient to say anything singular about his suffering. This dementalized mode of treatment espouses and encourages the impoverishment of the patient's aptitude to deal appropriately, that is, via symbolization and speech, with his depletion. The deficiency in question prevents him from integrating his symptoms with his psychic organization as a whole by associating them with his lived history. The virtual absence of dreams and phantasies is characteristic of such a deficiency, giving free rein to the patient's empty speech. This is translated by increased rationalization devoid of effect.

Is this operational thinking a consequence of the impoverishment of a society in the grips of external technological progress? Does it stem from a failure within the nuclear family to transmit to its children more adequate means for preserving their human integrity in the face of the uncertainties of their life? It is precisely such a colossal upheaval of our societies that we have been witnessing helplessly for the past thirty or so years. It is one thing to respond to the immediate needs of patients at the level of their suffering; it is quite another, however, to reduce their subjectivity and their mental functioning to these same symptoms.

Behaviourism confines itself to what it believes to be the *reality* of the human being, precisely where it is a question of not losing sight of his *possibility*. Taking the alienating manner in which individuals manage their suffering and pain for granted by forging a psychology that reduces them to their conditioned responses is more an issue of questionable ethics (Gori, 2011) than of a scientific approach.

Cognition, the corner stone of neuroscience

The term *cognitive psychology* was coined in a book published in 1966 by the American psychologist, Ulric Neisser (1966). However, notwithstanding the great success that the discipline in question has enjoyed, Neisser has distanced himself from it, criticizing it vigorously. According to him, cognitive psychology has strayed fundamentally from the essence of man by taking a linear model as its reference and by referring solely to laboratory studies (see Neisser, 1976, particularly the introduction to the first chapter). In particular, he adopted a critical position towards a book, well known by the public at large and co-authored by an American psychologist and an American political scientist, called *The Bell Curve* (see Herrnstein and Murray, 1996). According to them, ethnic differences substantially influence an individual's intelligence quotient (IQ). So they created a curve in the form of a bell on which they distributed, according to their criteria of intelligence, the population as a whole. We know that the scandalous welcome that the thesis of this book received among racist groupuscules. But over and beyond the opposition between the supporters and opponents of this assimilation between intelligence and "race", the question that arises is one of knowing whether the scientific postulates are really as ideologically innocent as they claim to be. It was no doubt in this way that cognition, in the formal and computational sense, put itself at the service of artificial intelligence, identifying the human mind with an automat. The inherent tendency of cognitive therapies to rely on the inclination of individuals to conform with well-established social norms was also maintained in this way, as if it were part of the scientificity they laid claim to.

American psychology, in general, is characterized by two major traits: *intelligence*, an object of predilection in its inquiries, and the *military initiative* in its research activities. A certain number of scientists, the clear minority, have sought to remedy this exclusivity of intelligence as an object of study. The work of Antonio Damasio, an eminent American researcher of Portuguese origin, is a case in point. His studies hark back, in their own way, to the philosophical thought of the old continent, snubbed by neuroscience since its advent. By opposing introspection as a method of research in the psychology of the nineteenth century, neuroscience refrained at the same time from questioning its own philosophical assumptions. In *Descartes' Error,* Damasio (1994) discusses in particular the history of an American railroad construction worker in the nineteenth century who survived a serious accident in which a tamping iron went right through his skull, damaging his prefrontal region. Damasio's question does not concern the cognitive aspect of the lesion, but rather the role it played in respect of social codes which were dramatically altered in the worker in question.

The exclusive study of cognitive phenomena is a feature of the physicalism of neuroscience, which is clearly embarrassed by the question of the phenomenal facts of the subject's experience such as colour, taste or vision. Indeed, these sensible qualities are unrepresentable and ineffable mental states. As such, they are neither observable nor quantifiable. These mental properties, known as *qualia*, are by definition unknowable entities for others, scarcely lending themselves to communication. They encompass a vast range of phenomena from the perception of sensible data (form, colour, sound, etc.) to diverse feelings such as pleasure or pain, including emotions and other entities experienced by the subject. Thus each author confers a different meaning to them.

There will be no difficulty in understanding why the qualia are so problematic for the cognitive sciences which function according to the fundamental rule of verification, that is to say, the verification of propositions reduced to the two logical categories of truth and falsehood, whereas qualia require neither refutation nor confirmation. The obstinate quest for objectivity is quite simply – as is well known – the reverse side of the subjectivity that it wants to combat. This is even truer when the domain of investigation concerns consciousness and individual experience.

The physicalism of the cognitive sciences has been the subject of criticism and refutations. This can be seen from the work of Max Bennett and Peter Hacker (2003), *Philosophical Foundations of Neuroscience*, considered as the first attempt to evaluate the cognitive neurosciences systematically. Mereology (taking the part for the whole, the brain for the person) is the main critique addressed to the researchers in the cognitive sciences by Peter Hacker, a philosopher of Wittgensteinian obedience and Max Bennett, the eminent Australian neuroscientist. This critique is especially judicious in that it does not only concern the research method of the sciences in question, but also their conception of the nature of the living in general and of man in particular. It sheds necessary light on the major question, since Descartes, of body/mind relations. Only the scientism of the end of the nineteenth century, which is still operative in many scientific domains, tried to make us believe that there could be scientific investigations independently of any conception about the way in which man conceives of the world. This relationship to the Real is at the heart of the psychoanalytic inquiry. We will see that it is also the touchstone of oneiric formation.

Vincent Descombes (see in particular Descombes, 1995) is one of the important figures involved in the critique of the cognitive sciences in France. His analyses, which also relate to Wittgensteinian thought, have acquired international scope and significance (see Descombes, 2011). He is close to the analytical philosophers of the Anglo-Saxon world whose position with regard to the neurosciences wavers between support and accompaniment.[1]

But above all mention should be made of the famous debate between Jean-Pierre Changeux, the neurobiologist of international renown, and Paul Riceour, one of the most eminent thinkers on the stage of French philosophy. To this day, their controversy remains the best work in terms of a critique of the neurosciences (Changeux and Riceour, 2000).

The emergence in France of the neurophenomenological movement in the 1980s by eminent researchers such as Jean Petitot (1993) and Francisco Varela (1996a) consisted, it is true, in trying to reconcile the cognitive sciences and

Husserlian phenomenology, but it also adopted a certain critical position towards cognitivism. It sought to naturalize phenomenology by taking up the early work of Husserl, particularly from the angle of mathematics which was supposed to orient cognitive research. But, at the same time, it refuted the two main theses of cognitivism, namely, computationism (equation of the brain with the algorithms of artificial intelligence) and representationalism (the reduction of every act of perception to the formal representations of the external world). Thus neurophenomenology finds itself in an awkward position both with regard to cognitivism and to Husserlian thought, which, for its part, rejects any physicalist approach to the mind. These theoretical divergences apart, the scientific contributions in neurophenomenology have proved very fruitful (see, among others, Petitot, 2003; Berthoz and Petit, 2006; Depraz, Varela and Vermersch, 2003). They have thus succeeded in pointing the way to a new orientation in the sciences of the mind, distinct from the functionalism governing the cognitive sciences. "We are thus justified," Jean Petitot (1983) writes,

> in saying that, structurally speaking, the formalization (in the naive formalist sense in which it is usually understood) is radically opposed to the mathematization which is, in itself, in conformity with the essence of things. There is an antinomy between the formal treatment of structures (whatever formalisms are used, from universal algebra to the theory of topoi, including the theory of automats) and their "mathematical physics".
>
> (p. 9)

J. Petitot's aim is to reach, by mathematical means, the objectivity of the Real in conformity with the method of transcendental philosophy which aims to return to things themselves.

Neurophenomenology was born within the prestigious Centre of Research in Applied Epistemology (CREA). It led to a split whereby it divided into two groups of researchers. The large majority of those who had a scientific training stayed at the CREA and the philosophers, of analytic obedience, set up the Jean Nicod Institute in 2002 – under the direction of François Recanati with the participation, notably, of Pierre Jacob, Joëlle Proust and Dan Sperber – with the aim of promoting the cognitive sciences as they are practised in Anglo-Saxon countries (see Feuerhahn, 2011).

My purpose is not, however, to take up *in extenso* the critiques mentioned above. I am more interested in concentrating on the behaviourist heritage of cognitive psychology concerning the stimulus/response model, notwithstanding all the opposition that it is said to have shown towards behaviourism. Certainly, cognitive psychology has succeeded in going beyond the behaviourist prohibition on opening the "black box" and to interest itself in the functioning of the encephalon; but it has not succeeded in dismissing the behaviourist conception of the putative model of conditioned responses. There are two main reasons for this continuity between the two disciplines.

The first consists in reducing all psychic activity to the basic concept which is *behaviour*, an English term which says more than its French equivalent, since it is the negation, purely and simply, of all subjectivity. Evidence of this can be found in the example of the Chinese Room. John Searle shows us by means of this fictitious example that the behaviour of a person who finds him/herself in an isolated room (behaviour which represents the artificial intelligence of automats) fools people outside into believing that he/she understands Chinese perfectly, whereas, in fact, the person is simply responding to the questions that he/she is asked in Chinese following a programme set up in advance. The person is doing no more, then, than executing this programme while respecting scrupulously the rules relating to it. But the execution of the programme, John Searle says, has nothing to do with understanding the rules that it contains. Thus (subjective) understanding is to be distinguished fundamentally from the behaviour that can be observed from the outside. Cognitive psychology was based on such a confusion.

The second reason stems – I have just given an illustration of it – from a fundamental conception of the living from which the first is derived. It concerns the dichotomy between the internal world of the living and its surrounding world.

In what follows we are first going to examine history and the pertinence of the model of conditioned responses before opening up the discussion on the splitting referred to above between the subject and the world. To do this, we are not only going to consider the question at the philosophical level but also its significance for ethology. We will then be in a position to speak of a current of thought that has remained dissident in neurophysiology and which lends support to my remarks on the specific temporality of the living. This will entail a primordial questioning which, I hope, will pave the way for understanding the oneiric world.

Note

1 Other than Ludwig Wittgenstein, already cited, philosophers such as Gilbert Ryle, Michael Dummett, Peter Geach and his wife Gertrude Ansecombe may be mentioned here from whom critics such as Hilary Putnam, Jerry Fodor, Michael Williams and Donald Davidson have drawn inspiration.

2

THE OBJECTS OF COGNITIVE NEUROSCIENCE

In this chapter we are going to examine different aspects of cognitive studies before tackling the question of movement which seems to me to be the true object of the physiology of the brain. This will allow us to clear the ground ahead before going more deeply into the various questions that are raised in relation to dream activity.

It will be a matter of finding a pertinent path that enables us to avoid the stumbling block of Cartesian dichotomy. Behaviourism would undoubtedly never have triumphed in the last century if it had not received its legitimacy from the Cartesian dualism between *res extensa* (body) and *res cogitans* (mind). This was how it dismissed the second in favour of the first. The brain as the seat of the mind was henceforth qualified as a black box and excluded from research. The only thing that counted was observable behaviour. And, even when, in the 1970s, cognitivism took over from behaviourism, the black box would retain its function in another form. Admittedly, it was now a question of opening it, but to give it a status close to *tabula rasa* by considering it as an interiority lacking stimuli from the external world.

Two principal concepts govern cognitive studies: conditioned reflexes and cognition. I have excluded from this list the question of the properties called *qualia* to which I have already referred. These are the subject of polemics and controversies within these sciences themselves and do not constitute, strictly speaking, their object of study. Daniel Dennett (2007, p. 75) once said that they were a poisoned gift for his discipline. The two scientific objects that I have chosen to discuss seem, on the contrary, to have an exclusive character for the cognitive sciences and, in this respect, condition their coherence both at the historical level and in terms of content.

Reflex theory

Reflex theory is one of the cornerstones of modern neurology. It is an integral part of neuroscience. Mechanistic thinking from Descartes to current-day cognitivism is constituted, in fact, by a current of thought dating back to the seventeenth and eighteenth centuries which gave rise to reflex theory. The paternity of the latter was, for a time, attributed to Descartes, an error since corrected by Georges Canguilhem (1955). A reflex, as we know, is an involuntary response of the organism to external stimuli. The circuit, called the *reflex arc*, starts from the afferent receptor neuron and reaches the efferent neuron after passing through the intermediate nerve cells. The nervous influx gets to the spinal cord without reaching the brain. In this way it served as a model giving birth to the behaviourism of a certain John Watson who excluded all interest taken in the brain. In the neurological research studies on the reflex arc, Watson could not fail to find a domain that was particularly promising for his stimulus/response theory.

In the history of the evolution of the concept of the reflex as a scientific object, Thomas Willis is a key figure. He was one of the greatest doctors of the seventeenth century, carrying out much painstaking research into the anatomy of the brain. In fact it was he who coined the term "neurology" for the science that he promoted, if not practically founded. We also owe to him the discovery of the extrapyramidal system backing up the vascular redundancies of the brain called the *Willis polygon* (or Circle of Willis). According to him, the striate body is the main part of the brain responsible for sensory perception. Though he was the first to employ the term "reflex" for involuntary actions, he did not succeed in discovering that the circuit of reflexes stopped at the level of the spinal cord. It is true that Willis had made significant advances in the understanding of the reflex (including its centripetal/centrifugal function), but it was primarily in the nineteenth century that the latter imposed itself as a scientific model and became completely integrated in the protocol of clinical examination (Canguilhem, 1955). This advance was possible thanks to the Czech physiologist, Georg Procháska, who treated it as a truly scientific object. Just as Newton had studied the force of gravity (*vis gravitans*) from its effects, without understanding its true nature, Procháska believed in the existence of an elementary energy, *vis nervosa,* in reflex activity, simply from its effects. Moreover, it is to him that we owe the expression *nervous conduction.* He was the one who explained how reflexes occured through the agency of the spinal cord, the rachidian bulb, and the central gray cores without the intervention of the cerebral mechanisms proper (Procháska, 1851). There then followed, in the history of the formation of the concept, eminent researchers such as Sherrington, Ramón y Cajal and, of course, Pavlov. Sherrington succeeded in establishing the integrative properties of the central nervous system, whose intervention is commanded by the reflex on each occasion. Here, as elsewhere, Sherrington gave a holistic explanation that ran counter to all fragmentary ventures in neurophysiological studies.

By the end of the nineteenth century, the reflex arc had practically become the explicatory model in neurology. Freud tried in turn to make use of it to explain

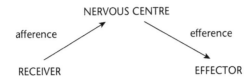

FIGURE 2.1 The reflex arc

FIGURE 2.2 The psychical apparatus according to Freud (*The Interpretation of Dreams*)

the hiatus between perception and the unconscious. A version of this can be found in the seventh chapter of *The Interpretation of Dreams* (1900). Having illustrated the psychical apparatus with the help of a schema beginning with the pole of perception and finishing with the pole of motility, Freud explains that "the psychical apparatus must be constitutive like a reflex apparatus. The reflex process also remains the model of all psychic functioning" (p. 538). His schema follows the same succession as the reflex arc, namely, afference-receptor-CNS-efference-effector, as shown in Figure 2.1.

Thus, like the system of reflexes, he concluded, the psychical apparatus is also directional. Indeed, his schema (see Figure 2.2) confirms that the apparatus in question only functions linearly and follows an order that is clearly determined in time. It requires no conscious understanding during perception.

The perceptual content (PC) enters along the afferent path and arrives at the muscular pole (*M*) without being apprehended consciously. The unconscious (*Ucs*) thus retains sensory traces (*s, s'*. . .) without a conscious system being necessary.

Freud's psychical apparatus conformed, both in form and essence, to the Pavlovian theory of the nervous system. At the 14th International Medical Congress that was held in 1903 in Madrid, Pavlov had defended the idea according to which the conditioned reflex was at the basis of the psychical apparatus, just like Ramón y Cajal, who, at the same Congress, established the neuron as the basic element of all study of mental activity. Having given up his neurological investigations, Freud did not attend the Congress but he continued to interest himself in the psychical apparatus according to other modalities in keeping with the discovery of the unconscious.

Three years later, in 1906, the Nobel Prize in physiology was awarded jointly to Ramón y Cajal and Camillo Golgi. This saw the beginning of neurobiology, the flourishing development of which finally gave rise to the neuronal man

(Changeux, 1983). During the Paris Congress in 1903, Cajal had already been awarded the Czar Prize, which Pavlov had had his eye on at the time. If we follow the path Pavlov took from the first Congress in Moscow in 1897 to that of Madrid in 1903, we can see the impressive evolution in his relationship to his work, one that was subsequently to prove decisive for the cognitive sciences (Campos-Bueno and Martin-Araguz, 2012). During this period, Pavlov, by virtue of his ambition, had succeeded in winning one of the international prizes of that era, thereby giving an unexpected extension to his theory of reflexes, promoting it to the rank of a promising basis for psychology. The most important moment in this evolution was no doubt the paper he delivered in French at the Paris Congress with the title "Experimental Therapy as a New and Extremely Fruitful Method of Physiological Investigation".[1]

Previously, Pavlov and his colleagues, Wolfson and Snarki, had only spoken, in their laboratory bulletin, of the psychic effect or excitation of reflexes, without equating them with subjective processes. Even if, subsequently, the physiology of reflexes evolved towards a more subtle integration, it has to be admitted that it has always retained its mechanical approach. It was this that came to constitute an inaugural approach to the study of behaviour.

It was in the years between the two Congresses of Moscow and Paris that Pavlov carried out what he had not been able to do hitherto, namely, experimental and systematic studies on conditioned reflexes. He thus crossed paths with Tolotchinov, who had already been conducting such experiments. Seeking to establish the mechanisms responsible for the psychic activity caused by reflexes, the studies in question finally turned into an experimental therapeutic approach, an old idea of Ivan Sechenov's.

Ivan Sechenov's work, which has largely been forgotten today, had a decisive influence on the future of reflexology, particularly with regard to the rapprochement between physiology and psychology (Dupont, 2011). Furthermore, Sechenov's scientific career attests to the tremendous expansion of neurology in the middle of the nineteenth century. It embodies the united research efforts underway at the time in the European research centres of Berlin, Paris, Vienna, and Moscow. The Russian intelligentsia, at this epoch, was carried along by an extraordinary momentum of what may be called Westernization, characterized by an interest in the singularity of the individual who was henceforth destined to assume his place within a society that was still feudal. Serfdom was, in fact, only abolished in Russia in 1861. It was no doubt this same quest for individuality that drove researchers to espouse the scientific ideal of the Western world. Thanks to his encounters with eminent European scientists such as Claude Bernard, Carl Ludwig, and Emile du Bois-Reymond, Sechenov was able to carry out his researches leading to the discoveries that left their mark on the history of neurology. It was not without reason that Pavlov called him the father of Russian neurology. This is borne out, moreover, by his theory of *central inhibition*. The encephalus was now endowed with a basic mechanism that allowed it to attenuate, and even to stop, the nervous influx of reflexes. The nervous system was thus not only governed by excitation, but also equipped with an inhibiting capacity to bind the force of the reflexes, thereby regulating the functioning of the organism. This discovery would be integrated later on in cybernetics, and Freud also put it to good use in his theory of repression. As for Pavlov, he made a behavioural translation of it, the consequences of which we are familiar with in American psychology.

Ivan Sechenov's work constitutes a conjunction between nascent individuality and the advance of European neurology. Notwithstanding his adhesion to idealistic theses in his youth, he was to develop a whole system of materialistic psychology on the basis of neurophysiological principles. "In my essay, 'Reflexes of the Brain'", he writes, "I developed the idea, from particular examples of increasing complexity, that it was possible to reduce all the main forms of psychical activity to the reflex type" (Sechenov, 1957, p. 209, translated from the French for this book). Thus began, at the end of the nineteenth century, the conjunction between the human soul and the neurology of reflexes, a conjunction that would give rise, barely a century later, to cognitive psychology.

The hazards and uncertainties of history saw to it that the work of another neurologist, the instigator of conditioned reflexes, took precedence over that of Ivan Sechenov. As a candidate for the Nobel Prize of 1902, Pavlov mobilized his laboratory team in St Petersburg with the aim of presenting the maximum of articles at the Congress of Naturalists and Physicians held in Helsinki in 1902. His personal contribution concerned the study of pepsin and rennet, two biochemical substances that he was far from mastering. After this mobilization had ended in failure, Pavlov returned to his conditioned reflexes, which he presented with great sucess at the Madrid Congress the following year.

The concept of cognition

I want to move on now to examine another scientific object of neuroscience, namely, cognition. Cognition is related to mental processes that are supposed to be measurable and objectifiable such as learning, perception, memory, decision-taking, intelligence or reasoning. The excessively vast nature of the concept – equating feelings and desires with the intellect and with discursive categories – reveals its provenance, that is to say, the logico-mathematical systems developed in the wake of the works of the positivist philosophers of the beginning of the twentieth century.

Cognition designates the process of an entity called mind, conceived of as a vacuity or empty space facing the external world. It is as if the entity in question only turned its attention towards the world in order to obtain the elements it needs from it, its function being that of processing the information it receives from the external world via its sense organs. Through the latter, the mind forges a *representation* of the world with a view to *processing* it according to its needs. By representation, one must understand the reduplication of the outside insofar as it is *re-presented* in the mind. Cognition is supposed to be the cerebral operation of a mental entity that, as a matter of principle, is divorced from the world to which it only has access through representation. It only achieves this thanks to the processing of information received from the outside by means of algorithmic devices (computation). These are operations realized from symbols based on logico-mathematical rules. The symbols in question are devoid of meaning; all that counts are their physical attributes. Their conformity with the external world is limited to what has already been deposited there. In this virtual reality that seeks to simulate the world, one clearly only harvests what has already been sown. The concept of simulation asserts that we only have access to the world by reproducing its formal properties.

On the contrary, "being disconnected from sensory input," Rodolfo Llinás insists,

> *is not* the normal operational mode of the brain, as we all know from childhood, when first we observed the behaviour of a deaf or blind person. But the exact opposite is equally untrue: the brain does not depend on continuous input from the external world to generate perception but only to modulate them contextually.
>
> (Llinás 2002, p. 7)

Some neuroscientists have tried to introduce nuances to this conception or to correct it. The work of Francisco Varela, a Chilean neurobiologist who died unexpectedly in 2002, is a good example. He contended, on the contrary, that the brain gives meanings and forms to the external world, while at the same time letting itself be modulated by it. Faced with the external world, the organism deploys its sensibilities. It opens itself to it while letting itself be inhabited by its properties. According to Varela, this "enaction" or coupling between the two entities – the subject and the world – creates each time a new coherent ensemble culminating in the existence of *emergent properties* (Varela et al., 1991). These properties come into existence thanks to the commerce, that is to say, the *path* of the subject in his intimate relations with the surrounding world.

We shall see later on that the concept of cognition in the sense of a vacuity depending on the sense organs is poles apart from the notion of *movement* as the intrinsic interaction of the brain with the world.

The view according to which the mind is no more than a *tabula rasa* dependent on sensory information has nonetheless remained intact in neuroscience, in spite of all the criticisms it has drawn from its own cognitivist philosophers. John Searle's *Chinese Room* argument is a case in point. However, it was in vain that, using this example, he tried to refute the widespread idea among cognitivists that the mind is a receptacle that only processes information put at its disposal.

The notion of cognition raises two main problems: on the one hand a restricted conception of the animal soul, the critique of which I have just outlined; and, on the other, the bias of neuroscience in favour of its primacy over other aptitudes of living beings. By focusing on the formal and abstract properties of cognition, cognitive psychology leaves aside entirely the content of sensible knowledge, relegating it to the void of a subjectivity whose existence it refuses to accept. It has had its place in the Anglo-Saxon philosophical tradition since Hume, a tradition in which the mind seems to boil down to a sum of impressions received along sensory paths. It is indeed by means of these impressions that cognition is reduced to perception, relegating other mental aptitudes to the background. What are the consequences of this primacy ascribed to perception? If a thing is given to me quite simply as an entity perceived in the external world, how can one explain that each time I find myself in front of just one of its aspects, I am able to grasp it in its totality? It is because the sensible always comprises what surpasses this visible presence. The sensible is, however, not given to me as such. It does not affect my sensible impressions, but it is immediately accessible to me. Is a thing an entity constructed by me or is it given to me as such by the world to which it first belongs? What is perceived is not only an entity immanent to the mind; it is just as much transcendent to it. During the course of this book, we will see that the subject/object dichotomy on which the notion of cognition rests cannot be sustained. My relationship to the world cannot be reduced to an intellectual and formal approach to what surrounds me. The supposed primacy of perception restricts, in fact, my commerce with the world by conceiving of it as no more than the sum of perceived objects.

Note

1 https://archive.org/stream/xiiiecongrsinter00inte/xiiiecongrsinter00inte_djvu.txt.

3
BODY/MIND RELATIONS

The body/mind split goes back to Cartesian thought which, consequently, gave rise to the separation between subject and object. Kant's *Copernican revolution* overturned this order of thought in order to go beyond the opposition between idealists (mentalists) and realists (physicalists). According to the first, the world is a product of the mind, whereas the second contend that there is nothing outside the external world. For Kant, access to the latter nonetheless remains dependent on the constitutive and *a priori* elements of the human mind such as time, space, and causality; indeed, man could never have access to the world if he did not approach it in an *a priori* manner. Jerry Fodor (1980) founded his *functionalism* on Kantian transcendentalism by assigning an innate and *a priori* language to mental processes. In his view, this language is conceived on the basis of a computationalist (algorithmic) model which claims to explain man's aptitude for gaining access to other, more complex models of thought. Jerry Fodor's Kantian model bears a certain similarity with the rules of *universal grammar* of Noam Chomsky (2011), the eminent American cognitivist linguist. Just like his teacher, Hilary Putnam, Fodor is one of the figures of *representationalism* known throughout the world; according to him perception and other entities of the mind are mental processes governed by a non-verbal but computational language which he calls *mentalese*.

Emmanuel Kant's transcendental philosophy found a very different expression in Edmund Husserl's phenomenology at the beginning of the twentieth century. *Intentionality* constitutes the first principle of Husserl's philosophical edifice. It signifies that any act on the subject's part is necessarily directed towards an object. Thinking is always thinking *about* something. Loving, hating, feeling . . . are directed from the outset towards the element that constitutes their immediate object. But we will see how psychoanalytic experience led Jacques Lacan to an even more judicious understanding of the subject/object relationship.

Before turning to the psychoanalytic position regarding the body/mind question, I want to mention briefly one of the most recent movements of cognitivist thought.

In the 1980s, we witnessed the institutionalization, within universities and laboratories, of the cognitive sciences. But notwithstanding the intense activity of research workers, cognitivist theory still seemed unable to fully support the hypothesis of such a functional identity between the brain and artificial intelligence. The various forms of progress in neuroscience created an increasingly great gap with cognitivist theory. The cognitive sciences began to be subject to criticism from the inside. This resulted in the elaboration of another approach called *connexionism,* which differs from cognitivism by virtue of its *absence* of programme. The automats conceived of on the basis of connexionist ideas are comprised of three "layers": the point of entry, the point of exit, and the intermediate connexions. Devoid of a pre-established programme, these connexions are supposed to stabilize and adapt themselves by readjusting their "synaptic weight" on each occasion, that is to say the gap separating their units called *formal neurons.* The stabilization of the automat (the computer) occurs in accordance with its contact with its environment. A new, but minority, current of connexionist thought has been trying recently to turn away from the obstinate refusal of its adepts to take into consideration the philosophical thought of the old continent. Drawing their inspiration from philosophers such as Husserl, Heidegger, and Merleau-Ponty, the authors of this current line of thought have not only contributed to bringing greater depth to connexionism, but also to promoting a certain number of models of brain functioning which henceforth distance themselves from the truism of cognitivism (see, among others, Varela et al., 1991). I will come back to this later when discussing the mechanisms of dreams.

The drive, a singular mode of the body/mind relationship

Is the drive, for Freud, a somatic force or a form of psychic energy? How, as a somatic force, does it succeed in transforming itself into psychic motion? Where does its creative or vital force come from?

According to Freud, the drive is at the frontier of the psychical and physical domains. What does "frontier" mean and what does the relationship governing the mental and the somatic consist of? The Freudian conception of *representation* attempts to answer this. "An instinct [*Trieb*]," Freud (1915b) remarks, "can never become the object of consciousness, only the idea [*Vorstellung*] that represents the instinct can" (p. 177). This Freudian notion of ideational representative (*Vorstellungsrepräsentanz*) has posed a certain number of problems for analysts, particularly in respect of its formulation. In Freud, representation in the philosophical sense of correspondence (*adequatio*) between subject and object turns into a notion of delegation (*Repräsentanz*). It is the representative of the drive in the domain of ideas (*Vorstellungen, représentations*) that is the mind. The question, for him, is one of knowing how the drive as a physical and endosomatic phenomenon manages to *enter* the psychic sphere. In other words, how is the drive transformed into an idea or presentation (*représentation*)? How does the drive, in its material "chemical or mechanical substance", Freud says, become a memory-trace in the unconscious? Such articulatons can only receive their impulsion, as we have seen, from a pre-established dualism between the psychic and physical domains. Representation as delegation is in fact what *corresponds* to the *drive*, which, for its part, is of a material order. But this correspondence is conceptualized on the basis of delegation. Cathexis, displacement, projection, inscription, denial, repression, and many other psychic mechanisms are all notions in psychoanalysis for naming the kind of correspondence (*adequatio,* for the scholastic philosophers) involved in the relations between the somatic and the psychic.

The expression "ideational representative" (*représentant-représentation*), it should be added, does not split the phenomenon of representation into two distinct parts. The first component (the "representative" [*représentant*]) has a *verbal* sense, whereas the second (ideational [*représentation*]) has a *nominal* sense. Being a material entity, the drive can only be represented (in the verbal sense of *delegation*) as an idea or presentation [*représentation*] (in the nominal sense of mental activity) in the psychic domain.

The drive as extended reality (*res extensa*) obviously cannot enter a domain that is in essence devoid of extension, that is, in the mind as *res cogitans*. Here we come up against the irreconcilable Cartesian dichotomy between these two substances (*res extensa* and *res cogitans*). Psychic reality is nonetheless dependent on its correspondence with objective reality, here the body. Delegation (*Repräsentanz*) is supposed to be the specific mode of such a correspondence.

Consciousness and representation

The external thing (*chose*) is related, in conformity with the psychological tradition, to *perception*. In other words, it has acquired the character of the *perceived*. The latter was transformed, with cybernetics, into *cognition*, and finally became the touchstone of neuroscience as a whole. The notion of *representation* underlies that of cognition. That is why several cognitivist authors consider that the Freudian theory of representation is derived from their own movement (see Erderlyi, 1985).[1]

But, unlike the cognitivists, Freud does not reduce perception to *consciousness*. In the *Project* (1950 [1895]), he establishes the distinction between perception and consciousness, assigning a secondary character to the latter.

A thing can be perceived, according to Freud, *without* being explicitly understood by consciousness. These remarks, which concern the topology of the psychical apparatus in the *Project*, assumed a more accomplished form in Chapter VII of *The Interpretation of Dreams*.

> We shall suppose that a system in the very front of the apparatus receives the perceptual stimuli but retains no trace of them and thus has no memory, while behind it there lies a second system which transforms the momentary excitations of the first system into permanent traces.
>
> (1900, p. 538)

This was how Freud came to the conclusion that "in the ψ-systems memory and the quality that characterizes consciousness are mutually exclusive" (p. 540). We saw above how, by making use of the model of the reflex arc, Freud tried to illustrate these same remarks.

At the theoretical level, the Freudian separation between perception and consciousness constitutes the fundamental condition of discovery of the unconscious. Henceforth, the supremacy of the subject of knowledge, the Cartesian *res cogitans*, is undermined. It is deprived of its central character of mastery assigned to it by the philosophical tradition. With Freud, the question of consciousness becomes obsolete.

If perception can reach the unconscious without being subject to the mastery of consciousness, the latter, in its secondary function, can only grasp it as an *after-effect of re-presentation*. The thing perceived may be said, therefore, to have an immediate relation with the unconscious; and if it can eventually enter consciousness, it is only in the second phase of the "*re*" of the *re*-presentation.

The constitution of the living

Is the body/mind debate a sterile controversy or a major source of impulsion? Do its premises rest on unsuspected presuppositions of modern subjectivity in which man seems to have related everything to his own existence? It will be useful, I think, to return now to the issue of the difference between animate and inanimate.

It will not be possible here, of course, to discuss all the conceptions that man has forged of himself as a living being; however, it is worth noting the concept of animation which was developed most fully in Aristotle's treatise *On the Soul*. Aristotle distinguishes the vegetative soul from the animal soul, assigning the nutritive and sensitive function to each respectively. The Latin *anima* (*anemos*) refers – as does the Greek *psyche*, the Arab *nafs*, or the Hebrew נְשָׁמָה, to breath. This conception, which is also attested to in Genesis, acquired its full significance in the schools of Alexandria, via Plotinus and many other thinkers, before finding its way to the doctors of the Middle Ages through the medium of learned Muslims. In modern times we have gradually witnessed its decline. It was in the twentieth century, both at the philosophical and scientific level, that it fell radically into disuse.

Descartes considered the living body as a machine. Excluding any notion of a sensitive or vegetative soul from it, he thought of it in terms of the mechanism of a "clock or other automaton" (Descartes, 1662, p. 113). For Descartes, therefore, the "organic organ functions without an organist". Quite to the contrary, Leibniz believed in a *prestablished* order in which the soul occupied a predominant place.

The idea of *organization* gradually began to take over from the *soul*. Although its origin goes back to Aristotle's treatise, the *Organon,* it was not until the seventeenth and eighteenth centuries that the conception of the living (*le vivant*) as organization flourished. Descartes, Bossuet, Pascal, Leibniz, and Kant each developed, in their own manner, the same organicist perspective. The *Critique of*

Judgement (Kant, 1791) states clearly that the organism is not simply the addition of a set of organs that work together to ensure the vital mechanisms of living entities; rather, each organ receives its function from the others and determines their functioning in turn. The concept that was later described by Claude Bernard as homeostasis translates this general organization on the physiological level. The animal was henceforth the unity resulting from this organic constellation. The biological philosophy of Auguste Comte, in the nineteenth century, was founded on what he calls *consensus,* a notion that names the *synergy* existing in *sensitive communication.* According to Comte, it is the latter that determines the intrinsic relation of the parts within the living organization. Comte (1996) applies this same concept in his social philosophy.

As a coherent and constant unity, organization is the concept with which science has sought, since the eighteenth century, to apprehend the essential mechanism of the functioning of living matter. We will see that such an infallible unity could prove to be an obstacle to such an understanding. A certain number of currents of thought would try in the twentieth century to remove this obstacle by introducing the concepts of the void and nothingness as constitutive of living entities (Mazauric, 1998).

The animal soul

The words *âme* (soul) and *animal* are etymologically identical. To establish a relationship between them, it is said, would be a tautological proposition. We will see that, in fact, this is not so. What is an animal? How does it differ from that which is inanimate? I have already outlined a certain number of conceptions that man has forged in this connection since the seventeenth century. Indeed, it was from the moment that man became the source of his own questionings that the question concerning life also began to preoccupy him (Foucault, 1966). The different conceptions about life in general and about animals in particular seem to approach the animality of animals from an external viewpoint. Man nonetheless shares his condition with animals. What are we, then, insofar as we are animals and what do we share with them? It is important to avoid falling into the trap of indulging in inconsequential introspection at this point. Can animals be considered as subjects? What specific experience gives an animal access to its being? We have just designated an aspect of an animal that seems to belong to it, namely, that of having, as it were, access to its own being in the form of feelings, whether painful or pleasant.

By *access*, I mean the relation man has with himself without resorting to any conscious or rational reflection. To use a philosophical term, it is a *prereflexive* and not a thematic relation, a relation that is the obverse of the Cartesian *cogito*. This relation stems, quite evidently, from the first "opening" which *makes* man a "clearing" (Heidegger) that receives being as it offers itself to him. It goes without saying that without prior access to himself, his openness to the world could not exist. But would it not be incongruous to argue the same thing in respect of animals?

The access I have just been referring to depends on a more fundamental element. It implies a sort of mismatching (*écart*) at the very heart of the existence of the living, which is the support for such an access. If we regard man as an indivisible totality, we run the risk of eclipsing his primary condition. Indeed, he stands constantly in a double register in relation to himself: he encumbers himself with his existence as much as he seeks to evade himself. He tolerates his being as much as he finds it intolerable. Can the same be said of animals? Is an animal also mismatched with its being? This is where an animal differs from a machine, from an automaton. A machine is at one with what it produces. Can we always obtain from animals, as is the case with machines, the same invariable response without a mismatch of any sort? Science, nowadays, is based on such a postulate, with unforeseeable consequences. The fact that we are able to obtain under laboratory conditions the same response to the same stimuli, that is to say, with no variation, does not give us the right to equate animals with inanimate matter, even an intelligent machine. Confronted with pain or danger, an animal reacts. Is it the animal that reacts or is there something in it that reacts? An animal's reaction to danger and need counts among the phenomena that led man to forge the concept of instinct. Why does an animal react to danger? Answer: to preserve its existence. Does that amount to saying that an animal possesses, in one way or another, a sense of *its* survival, a prior access to its annihilation? In other words, does death – which runs counter to what it tries to preserve – exist *a priori* for an animal? If not, how could it react to danger? An animal's relation to pain and danger is of such complexity that I can scarcely go into it here. The male praying mantis lets itself be devoured by the female after mating. In male ticks, death also occurs after mating with the female. When an antelope is trapped by lionesses, it *seems* entirely resigned to its lot. Does it let itself die "willingly"? Or is it paralyzed by fear? If so, fear of what? Is it overwhelmed by fear due to the presence of its predator or does it succumb to the fatal outcome that it already senses or expects?

It is as if in animals there is a mismatch both in relation to the external world and to these factors of an internal order to which Charles Otis Whitman, at the end of the nineteenth century, gave the term *endogenous movements*. Whitman noticed that within the same category of birds, a certain number of behaviours differed from one species to another, distinguishing them from each other as much as their morphological characteristics. It was these specific behaviours that he called *endogenous*. He speaks of them as if these birds had acquired a sort of "social ego" (Wilson, 1975).

The *lure technique* implies that any stimulus can trigger the same response provided that it has the same *form* and presents the same image to the animal. The latter is misled and consequently manifests the same response as if it were a real stimulus in keeping with its endogenous behaviour. The phenomenon of the lure reaches its zenith when employed by the animal itself. Think of lionesses using ruse and cunning to trick their prey. Depending on the wind direction, one of the lionesses starts hunting at one end of the herd, while other members of the group wait for their prey at the opposite end. Alerted by the smell, carried by the wind, of the first lioness, the herd takes flight in the opposite direction where the other predators are waiting for it. We are familiar with the decisive role that the image plays in the animal kingdom; and we know, too, that the lure and the image are essential elements in the access of the young human being to the mirror stage, giving rise to his *ego formation* when he begins to distinguish himself from his fellow human beings. The symmetrical relation of the image phenomenon is dependent on this kind of mismatching that I am trying to examine in connection with animals. It is this same mirror-relation, dependent on the image, that is responsible for the triggering of two major phenomena in man and animals, namely, aggressiveness and sexual behaviour. The dual trigger of aggressiveness and competition leads a bird to steal the nest of another bird. It then immediately ceases to build its own. This is where animals differ fundamentally from machines. As early as 1895, Groos contended that the playful activity of animals could be more easily explained in terms of the phenomenon of *imitation*, that is to say of image, than in terms of an excess of energy. Later, in 1956, Lorenz gathered together the observations necessary to support the thesis that clearly distinguished playful activity from excessive activities. He noticed that during playful exchanges with a cat or a dog, our brutality does not make the animal adopt a fighting posture (Lorenz, 1978). The animal knows that it is a game.

Lorenz's discovery in ethology, in 1935, of the phenomenon called imprinting (*Prägung*) gave even greater consistency to the relation between animals and the image (see Lorenz, 1970). Heinroth had already noticed that geese adopt as their parent the first large mobile entity that they meet at birth. The "adopted parent" will be considered as a real member of the species and, even in adulthood, will constitute a sexual object of predilection. It can be seen, then, that the first image, charged with affect, that emerges after birth can trigger the process called *impregnation or imprinting*, provided that it represents a certain configuration. Hence the theory of attachment promoted by Bowlby (1969, 1973, 1980), the eminent British psychoanalyst, who, following the investigations of Konrad Lorenz in ethology, confirmed the studies of René Spitz (1945) concerning children who were victims of hospitalism in the period after the Second World War.

Different types of mismatching constitutive of the living

At what point does it become possible to speak about *mismatching* in the animal kingdom? Admittedly, mammals are more apt than other species to have access to such a relation differentiating them as much from themselves as from inanimate beings. We do not know the exact nature of such a gap, and even less the manner in which animals experience it. The sad spectacle of animals kept in cages or in captivity backs up this idea of mismatching. Is it just a meaningless impression stemming from our projection on animals? Are sadness and joy among the categories that do not exist in animals? Do we attribute the latter with the characteristics that we find in our pets? Lacan (1974, p. 16) coined the neologism "*hommestiques*" to suggest the extent to which domestic animals could approximate to the world of humans endowed with language. In any case, *feeling* is not reducible either to sensation or to perception. *It is the relationship that the animal has with its own sensation.* It is precisely this relationship that I am going to call *mismatching.* It goes without saying that we cannot take feeling for consciousness, even if it is a necessary step towards it. Consciousness can occur at any moment in animal evolution; it does not, however, constititute a condition of it.

I will qualify the primary gap or mismatching in man by the term facticity (*Faktizität*), which refers to the human aptitude to have access to his being as his *irreducible condition.* Man has constantly thought of himself as having been thrown into the world, whether in the form of dereliction or in the form of a reality principle. It was while thinking about this irreducibility that Sartre defined, in his own way, the concept of facticity. "The fact of not being able not to be free," he writes, "is the *facticity* of freedom" (Sartre, 1943, p. 567). Facticity is thus the primary condition of man as a living being. It refers to this first distancing, however vague and ungraspable it may be, with his own being. The gap, in the sense of facticity, is the primordial outline of the living being insofar as he is capable of turning round, however slightly and summarily this may be, towards himself.

Mimetism is another mode of mismatching. This is followed by the ego symmetry that constitutes the so-called mirror stage in the human infant, and then by language, which radically changes the realm of the living and permits man to distance himself from his animal condition.

Consciousness, understanding, reason, logos . . . are all terms that have had diverse fortunes in the course of the history of human thought. These terms were used to designate the highest rank of mismatching practised by the speaking being. The advent of psychoanalysis changed the situation completely. The human condition underwent a fundamental decentering. Man was no longer master of himself. This designates the specific dimension of man that Lacan calls the division of the subject ($, i.e. *barred subject*), the highest level of mismatching in oneself that can exist. I will come back to this later.

The "*objet petit-âme*", what is closest and yet most distant

What object is man for himself? What links him to his being considered as most truly his own? What do we mean by ownership in the relation that links the human being with himself? Let us put aside right away the illusion that consists in thinking about this relation in terms of completeness. However sustained narcissism may be in an individual, it is never completeness. We will see later how the human constitution strives against such a feeling. The object that is the human being for himself is diametrically opposed to completeness. For reasons of ease, I am going to refer to it by the term, which is loaded with a long and heavy history, of *soul* (*âme*). It is not my intention, however, to awaken the old demon that inhabits this fictitious entity, conceived of since Antiquity as a substance. Likening it to *narcissistic love*, Lacan (1975a, p. 84) conjugated it in the present tense: "I soulove myself, you soulove yourself, he souloves himself . . ." (*je m'âme, tu t'âmes, il s'âme . . .*). Still called today *anima* in Italian, soul (*âme*) goes back in fact to *anima* (tenth century), *aneme* (eleventh century), from which *anme*, and finally *âme* (see Bloch and Von Wartburg, 2008) is derived. The notion of *animal soul* seeks to name, by its very redundancy, the specific relation that an animal possesses within it with that by virtue of which it is alive.

This brings us back, then, to *mismatching* as a distinctive trait of the living. The *identity* of the soul and of the *animal* is in fact due to their mismatching. The question is: how does the soul "soulove itself" (*s'âme*), and in what form does it relate to itself? There is a counterpart in the human subject's relation to the other that Lacan designates under the name *objet* (petit)*"a"*.[2] This object names the *other* as similar to me. Given its "similitude" with the "âme animal", there is justification for calling it the *objet petit-âme*.

The *objet "a"*, Lacan says, is *unrepresentable and non-specular*. This is a definition that enables us to get beyond the difficulty we had of designating the specific mode of relationship that the animal, man, has with himself. That is why, in speaking of the *objet "a"*, Lacan says: it is that of which we have *no idea*. The object in question does not exist in the strict sense of the term (it is not an entity or substance). At the same time, Lacan says, it is "something". It would be the animal soul, if it existed. In other words, there exists *qua* nothingness, a nothingness that has not been nihilated, but which is *established* as nothingness. That is to say, it only exists in the form of void. Its existence would have satisfied the subject so much that he would have been annihilated as *subject*. This is because, for the psychoanalyst, the object is not an *effect* of the subject. On the contrary, it is the *cause*. It is the object that causes the subject's desire and relaunches it over and over again. Being ungraspable, the *objet "a"* only appears in the form of its *éclats*, in the two senses of fragments and brilliance. But at the same time it is the first object, *lost* forever, Freud says. The subject no doubt places more value on it than he does on himself.

Let me illustrate these remarks with an example. Eva is a young girl of thirteen who is afraid of getting into her father's car. Six years ago her father had a serious accident but quite miraculously escaped unhurt. That day, by chance, he was alone at the wheel. Eva often dreams that she is at the wheel of this car. On the back seat, there is a little girl: Eva herself, at the age of 7. The car ends up crashing at the bottom of a steep hill. Eva has lost control of it.

In this extraordinary enactment, where is the subject and what is the object? Who is saved and who dies? Is Eva at the wheel or on the back seat? What trauma is involved? Eva was not in the car at the time of the accident. *But she could very well have been in it.* The trauma constituted itself for want of presence. The *objet "a"* is precisely what exists for want of presence; it is a hollow in which the *division* of the subject is constituted. In this respect, it is the most primordial trauma of man. What the young girl is holding on to in her phobia, what she wants to preserve, what is dearest to her, is precisely the little girl that she still is for her father, the immensely precious *objet "a"* on the back seat of the car, which has evaporated with time, but which persists in its lack of existence as a *lost object*. Eva *souloves herself* (*s'âme*) just as her father *souloves her (l'âme)* as the little girl that she is no longer. The little girl is an *éclat* of her *objet "a"* even though the latter cannot be reduced to it. For Eva this *éclat* is in fact what *remains* of the *failed encounter* between her and her father, between her and the nothingness which could have carried them both away in the accident. The *remainder* is one of the essential functions of the *objet "a"*. This residue is what *remains* of the lost object of which Freud speaks. We know that what has been lost has never been reached, for the object is always lost. It never ceases to lose itself again; hence the repetition, that is to say the constancy in which my own being *continues* as always-already lost. Of the lost object, the subject will have nothing but fragments in the form of partial drives, which he takes each time for his *objet "a"*. It is here, indeed, that we can really understand the relevance of the Freudian conception of the drive.

The following are some of the fundamental characteristics of the *objet "a"*: its characteristic of *residue* as a reminiscence of the lost object which confers on it its always *partial* aspect and its *éclat*[3] which abducts the subject from himself and henceforth causes his desire.

By means of the partial object, the subject tries to represent to himself the *objet "a"* which is none the less unrepresentable. The *objet "a"* can, in effect, embrace the partial drives without reducing itself to them. The concept of drive in Freud is a bridge thrown down between body and mind, between the somatic and the psychic. Freud assigns it with the character of representation. The latter suffers, as I have already said, from an ambiguity. Representation sometimes designates the activity of forming an image and sometimes the act of delegation. I have interpreted the Freudian notion of *Vorstellungsrepräsentanz* in the following way. The drive is what represents (*delegates*) the somatic for the psychic. As soon as the somatic entity is represented at the psychic level, it takes on the status of representation. Similarly, the *objet "a"* espouses the drive and *opens up* a path for itself at the heart of the representations without being reduced to them. As Lacan says, it can neither be represented nor espouse the specular image. It is what never ceases to be absent from its place. Thus it is mismatching, *par excellence*, a gap, which, as we will see later, pertains to the real (*le réel*). Being the failed encounter, the real constitutes itself as trauma. Eva's dream and her failed encounter with death, where the *objet petit-âme* merely reveals her abyssal being, attests to this.

The constitution of the subject

Faced with the sterility of the body/mind debate, which seems to have lost its questioning force, I turned towards what distinguishes the animate from the inanimate. This in turn led me to the question of *mismatching*, a division that marks the living by bringing it intrinsically face to face with itself.

The living carries its being itself. It is by virtue of this gap that it welcomes the world within it, without it being possible to separate the world from it and to reduce it to an inconsequential interiority. However, the concept of cognition clearly designates the contrary. According to this conception, the mind, as a closed entity, only turns its attention towards the world in order to inquire into what it needs from it; the world is thus conceived as a place of information and from a pragmatic point of view. In other words, from this perspective, the living only opens itself to the world to extract data that are useful for processing mentally.

For empiricist philosophers, who consider that the external world is the sole entity worthy of research, the mind is first and foremost an aptitude for representing it. It was in the face of such empiricist *realism* that Kant carried out his Copernican revolution by demonstrating the impossibility of apprehending the world without *a priori* conditions of understanding. It is thanks to these categories *prior* to sensible experience that man is able to have access to the world. Kant's transcendental philosophy not only repudiated the realism in question, but also its opposite, the idealism that claimed that everything is dependent on the human mind.

With Kant the question subject/object was henceforth posed in other terms. Husserl's phenomenology went a step further at the beginning of the twentieth century by doing without *a priori* categories and by positing man's openness to the world as *constitutive* of his consciousness. According to the principle of *intentionality*, every act is always-already related to an object. In other words, all consciousness is always consciousness *of* something. Our thoughts, our perceptions, and our feelings cannot exist without being related intrinsically to their objects. The constitution of the object occurs in the very *act* of the subject.

The advent of the Freudian discovery of the *unconscious* brought an even more decisive response to the subject/object question. Freud himself scarcely employed the term "subject"; it is the term "object" that appears more often in his writings. It was Lacan who taught us to identify the place of the subject in his intimate relation with the object. But in my view Lacan was right in affirming that this *intimacy* refers to nothing other than his *"extimacy"*. For the subject constitutes himself on the basis of a relation of alterity towards himself; this is what is called *subjective division*. This shows just how much the object *divides* the subject by making him "ex-sist". In other words, the Lacanian subject is neither the agent nor the ". . . living substratum needed by this subjective phenomenon, nor any sort of substance, nor any being possessing knowledge in his *pathos*, his suffering, whether primary or secondary, nor even some incarnated logos" (Lacan, 1973a, p. 126). He is rather the *effect* of his act, an act that is none other than the desire that divides him and distances him from himself. Strictly speaking, what does this act consist of? Answer: it *subverts* the subject by placing him in a conflictual relationship of desire, engendered by the object. Division designates the subject as the effect of his object, which causes his desire. That is how Lacan distinguishes the subject from the subjective. The latter is what is contained within the entity called mind. It is by believing in the subjective that the prevailing reductionism leads the cognitive sciences astray, encouraging them to think of man in terms of cognition, that is, as an entity emerging from time to time from his subjective shell to ask for external information.

Let us return now to the Lacanian subject in his intrinsic relation to the *objet "a"*. For Lacan, the subject, as I have said, is neither substance nor agent; rather, he is the effect of the object which causes him. Conseqently, the subject only constitutes himself insofar as he is divided. Division means evanescence, fading, or subversion. The subject is only instituted by what links him to his object, that is, to his desire.

Lacanian thought concerning the subject claims to be non-ontological, rejecting every category of being. The subject, for Lacan, only ex-sists in relation to his object. This ex-sistence is effected by virtue of language, a treasure of *signifiers*. Language is the *locus* of the *Other,* that is to say of the alterity just mentioned. The subject can only signify himself by referring himself perpetually to another signifier. This is what leads Lacan to state that the *signifier only represents the subject for another signifier* (Lacan, 1966c, p. 694). The representaion involved in Lacan's formula is to be understood in the sense of delegation. It does not rest on any kind of representationalism. The assertion that "the signifier is what represents the subject for another signifier" is illustrated by the topological figure of the *Moebius strip* (see Figure 3.1).

As can be seen, this strip is without a top side and underside. And yet each of its "two" sides designates a different signifier of which the torsion point separating them is no other than the subject. The subject is the cut (the division) that is referred from one signifier to the other. This means that the subject is subject to the chain of signifiers without succeeding, however, in discovering the one that can really signify him and help him overcome his division.

The subject is never this or that. To qualify him in this way would be to reduce him to what is most foreign to his essence. He is represented by a multitude of signifiers which refer him, each time, to other signifiers. To reduce him to this or that would simply have the effect of going against his constitution. This is why all racist discourse is devoid of meaning.

In the act, too, the subject only determines himself retroactively. In other words, he finds himself transformed by his own act. We can thus say with Lacan that the subject *identifies himself* with his act. It is here precisely that he constitutes his object. This ex-sistence designates a mode of temporality in the subject

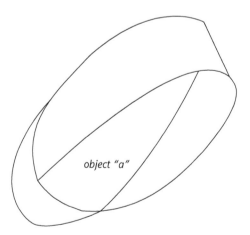

FIGURE 3.1 Moebius strip

which deserves closer examination. But before doing so, I want to turn my attention to the median void of the Moebius strip, the locus of the drive-based object (*objet "a"*).

The drive in Freud and Lacan

The drive is usually presented on a formal plane. Hence the well known schema that Lacan gives of it, which I will be discussing later on. I have already mentioned the link that Freud establishes, with the help of the notion of the drive, between the body and the mind. But at the clinical level, the drive designates, first and foremost, the impulses, both somatic and unconscious, which, like veritable earthquakes, permeate the subject's affective life. The drive is considered at once as temporary, during the space of a satisfaction, and indestructible, always ready to go beyond repression. It searches obstinately for its object. It is nonetheless capable of admitting other destinations, and even of transforming itself into its opposite. Dream formation seems to be based on a distinctive mode which I shall qualify, in my investigations into dream life, as the reverse of the drive.

In the light of these considerations, in what follows I am going to leave bibliographical references to the work of Freud and Lacan aside in order to present their respective articulations as a whole. This seems to be the best way of proceeding if we want to enter into a dialogue with their modes of thought and integrate in a personal way their respective ways of thinking about the drive.

The world of the newborn infant is characterized by what may be called, with Lacan, "the hum of the pleasure principle". We need here to picture the instinctual life of the baby as closely as possible if we want to understand his specific mode of being-in-the-world. The drive constitutes the infant's relation to the vital elements of his world, that is to say, what we usually refer to as its objects. According to Freud the object is that "little thing that can be detached from the body". It is the element that "separates itself" from the body, such as the maternal breast, the baby's thumb, its tongue, or any other part of the body that is prone to accumulate tension. This accumulation, in whatever part of the body, can constitute a centre of tension which Freud calls an *erogenous zone*. The erogenous zones are little islands cathected with tension that can be localized in different parts of the body. How are they determined? How does tension become localized in one part of the body and not another? What is the exact mechanism of the relation that is established between the subject, his erogenous zones, and the objects that he is led to cathect? The answer is to be found in the articulation between need and demand. *Need* can be given a biological status which is immediately caught up in the network of *demand*. Here, it is a question of the demand of the Other; in the example of the infant, it is the demand of the mother. The mother, the infant's primordial Other, transforms her own desire into the infant's demand. She receives her desire in the inverted form of the infant's. In other words, her desire is disguised as a demand coming from the infant. The infant/demand is in fact the mother/desire: the statement "I desire this" is substituted by "You are demanding that". In this way, biological needs are from the outset alienated from their basic biological nature and inscribed within the complex of the human relations of desire and demand. In the infant's demand, the mother sets up her own being. She gives form and object to his demand. In the act, for instance, of breast feeding, the mouth becomes a zone of tension and the breast its object. Thus the object is the *element separated from the body by the desire of the Other.* The breast/object is determined here by the mother's desire.

Why do we call the breast an "object separated from the body"? The answer lies in the particularity of the drive. The latter is an act without a real object; for, if there is an object, its reality is not cathected but circumvented. Take the example of sucking an object like the thumb. The infant derives no "real" benefit from the thumb. The impulse of sucking is the circular movement of the mouth around the object/thumb. The latter is circumvented in the act of the drive; it is not incorporated, obtained, or grasped by the baby. It constitutes the "zero" object which is only there so that the repetitive movement of sucking can take place. By object I mean the "nothing" that makes repetition possible. The latter exists insofar as the object re-emerges from its nothingness during each cycle of the drive movement. The object merely sustains the repetition of the act. If it existed in its "reality", if it was "full", the act of the drive could not exist. Pleasure is only obtained as the effect of this act and not as the appropriation of the object. The drive creates a void in the fullness of the object. If the object of the drive was full (real), the latter would be sated and devoid of its nature. And so, the drive remains insatiable; this is what determines its essentially repetitive character. An animal sates its hunger. Only man can eat in excess. Anorexia, and its corollary bulimia, do not exist in animals. Eating is a psychic act.

If the object of the drive is "empty" with its fullness, if it is only circumvented by the pressure of the drive, it does not on the other hand cease to constitute itself within the body. *The object is the negative of the body* on the outside just as the body is the negative of the object on the inside. The circular movement of the drive forms a complete circle divided into two semi-circles. One of these circumvents the real object on the outside of the body and the other circumvents the "hallucinated" object on the inside. Between the two semi-circles is the surface of the body, that is, in general, the orifices. This surface is a rim without any thickness. The *pressure* of the drive originates from this surface, called *source*, and follows its trajectory which consists in circumventing the object in order to reach its *aim*, that is to say, the discharge of tension.

These four elements of the drive, the source, the pressure, the trajectory, and the aim are in play twice over: once on the semi-circle on the outside around the circumvented real object and once on the semi-circle on the inside, around the "hallucinated" object (see Figure 3.2).

As the division of the circular movement of the drive into two semi-circles, situated diachronically one after the other, is purely a didactic artifice, it is important to add that these two so-called movements occur at the same time and synchronically. In other words, not only does the pressure of the drive circumvent its object but in addition, and at the same time, it sees to it that the body constitutes itself as an object.

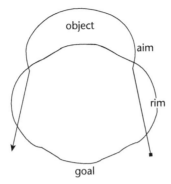

FIGURE 3.2 Partial drive

In the oral drive, for example, in the act of chewing gum or thumb-sucking, the body constitutes itself as the object of the drive, here the gum or the thumb. The body makes itself the negative of the object, sucking becomes the equivalent of being sucked and vice-versa. As Freud said, in the act of kissing, it is the mouth that kisses itself. We can see here that the subject in the drive is himself in a state of *perdition*; he becomes its object, he makes himself object. In the oral drive, for example, he makes himself breast, or in the anal drive he makes himself excrement. This is the truly auto-erotic character of the drive: the subject has become object, but this object is just a vacant place that can be "occupied" by any object. For example, in sucking, it is of little importance whether the object is the infant's thumb or an object such as a toy. What matters is the back-and-forth movement of repetition, which obviously does not reach its final aim (the satisfaction of a need), but produces pleasure. This *effect of pleasure* is a bonus, for it does not constitute the sought-after aim, namely satisfaction. Here we are faced with a paradox. The drive movement is triggered with the aim of reducing tension. At the same time, the latter is gradually increased by the repetition of the drive itself. The drive diminishes the tension, resulting in the effect of pleasure; it also increases it, which is why repetition is a factor of tension.

This paradox embodies the contradictory nature of the human being, that is to say, ultimately, the intertwining of the life drives and death drives. The repetitive and mechanical character of the drive tends towards the erosion and even the annihilation of the libido, even though it is triggered, precisely, in order to support the life drive. A singular illustration of this can be found in psychosomatic illnesses.

One of the distinctive traits of the drive in the newborn, in contrast with what one finds in the older child or in the adult, lies in the Freudian concept of the pleasure principle and the reality principle. To take up Lacan's articulations, it must be said that the relation between pleasure and reality is of a singular kind. Like the particular topology of the *Klein Bottle* (see Figure 3.3), the baby's body, intertwined as it is with the external world, is a volume whose surface – which extends from the inside (pleasure) towards the outside (reality) – suffers no separating cut between these two entities. Thus the transition from pleasure to reality, or the converse process, depends on a uniform space without differentiation between the inside and the outside. It can be seen, then, that the supposed dichotomy or opposition between the internal world and the external world no longer holds as soon as we do justice to the experience of the subject.

The drive is closely articulated with the demand of the Other. It would like to create its own demand. The subject of the drive is a solitary subject who, in the absence of the Other and of his demand, fabricates a world of similitude in which he is at once subject and object, the one who *demands* and the one who is *demanded.* Contrary to desire, in which the subject is divided and constantly exposed by its gaping hole, the drive is an act in relation to which the subject, with all due respect to the cognitivists, is headless (*acéphal*). He is a lacunary apparatus, his object being nothing but a void. The relation between desire and the drive is governed by the fact that the drive is *the bodily seat of desire.* The danger in the drive is where it is no longer in a position to respond to the call of the Other of desire. Here, the drive would like to escape from the sway of the Other. It is the permanent temptation to efface the latter with the aim of substituting itself for it.

The death drive is operative when the subject is no longer subjected to the desire of the Other, the very desire that divides him as a subject. As it has the function of supporting desire somatically, the drive should remain closely linked to it. Let us say that the act of the drive is the lure in which the desiring dimension of the subject engages itself. The lure according to which it is possible to overcome the lack inherent in the object. The act of the drive is the temptation to forget this gaping hole. The Other of desire thus becomes the other (object) of the drive. It constitutes itself as the fragmented and partial object. It is no longer the mother as the Other determining desire, but rather the breast as other, that is

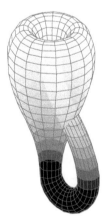

FIGURE 3.3 Klein bottle

to say, as a partial object. Consequently, the breast (*objet petit "a"*) is cathected to the detriment of the mother (the Other). The subject, actively listening to the desire of the Other, transforms himself into a drive-based subject denuded of subjectivity, for he is a headless subject (Lacan, 1973a, p. 181). Desire and its somatic side, namely, the drive, must remain closely articulated with each other. Otherwise, pure desire remains an inhibited impetus without a somatic basis and pure drive a movement towards death. This is exemplified by the autistic child whose only place in the desire of the Other is that of an abyssal vacuity. He has not, so to speak, *arrived in the desire of the Other*, an Other who could give him everything, care, empathy, and even love and sacrifice, except a *place* of desire.

* * *

Among the countless objects capable of occupying the vacant place of the object of the drive, psychoanalysis has identified four that are considered as the most essential: the breast, excrement, the gaze, and the voice. As far as the invocatory drive is concerned, the voice constitutes the object detached from the body that allows man to "make himself voice" ("*se faire voix*") so that he can reach the ear of the Other. I make myself voice so that it comes back to me from the Other. It is with my voice that I invoke the Other. The circular movement of the drive is thus closed again.

The invocatory drive in the infant could be the subject of a separate study. Here I am just focusing on one of its crucial moments in the mother/child relationship, a primary location of the interweaving between drive and desire.

The mother's singing, commonly called the lullaby, is one of the special moments in the relationship that forms a close bond between the infant and his mother. The lullaby is a sign of the mother's presence. The infant finds it reassuring and soothing. The rhythm of the mother's voice procures for the child the to-and-fro movement that is indispensable to the drive. The infant makes himself voice, the voice of the mother, and his being dances to the rhythm of her singing. Here the life drive is at its height, carrying the infant into a state of pure absence, akin to the death drive, making it possible for him to go to sleep and withdraw his interest from the external world. What is there in the mother's voice? What is the extraordinary charm in it that allows it to transform an agitated being into such a peaceful body? If going to sleep is a drive-based act, where is its source? On what erogenous zone is it based? The infant makes himself voice; he is all ears. The gaping hole of his listening extends over the whole of his body. Here it is not a question of a partial erogenous zone, of an isolated orifice; the baby's body is entirely immersed in the singing and the voice of the mother. The modulations of her singing constitute the circular movement of the drive.

The lullaby has a particular timbre, a special rhythm, and a significant tonality. It does not often suggest great joy, but it is not dull. One is overcome by a chrystalline limpidity, but one that is undoubtedly charged with sorrow. Why is there this almost elegiac tonality in which a complaisant melancholy emerges as in the time of the troubadours? If that is true, how can such singing, coloured with sadness, soothe the infant so deeply?

The link between pain and loss is undeniable. If the lullaby is coloured with sorrow, what loss is it evoking? Loss or mourning, mourning and death, these are the things that we are most familiar with.

Between the mother's lullaby and the infant's falling to sleep, the tuning is so perfect that their "fusion" remains faultless. This completeness excludes any third party and considers the latter as its true enemy. The lullaby baby reveals the quarrel that goes back to the time when the father was put to death so as to exist eternally in human guilt, the latter being the recognition of the debt that men contracted unwittingly through the murder of the father. Indeed, one of the mother's unconscious phantasies conceals the wish to abduct the child from the father.

This guilty wish makes possible the fusion between mother and infant and their mutual investment that is so highly charged with libido. As she is singing, the mother is under the illusion that her fusion with the infant is eternal. This phantasy thus highlights the ambiguity of her feelings of love and hate towards the father, which are at their height here. Parricide (the unconscious phantasy of eliminating the father) is constitutive of the law of incest. The civilization of human societies remains dependent on it (see Lévi-Strauss, 1948).

Notes

1 See also those authors who retrace the common philosophical origins of cognitivism and psychoanalysis (Dreyfus, 1982).
2 To distinguish between "objet *petit* a" (the other) and "*grand* A" which stands for *l'Autre* (the Other).
3 In the very sense of the Greek *agalma*.

4

QUESTIONING TEMPORALITY

I am sitting behind my desk and get up to go to fetch a book from the other end of the room in my library. As a subject, I "carry" my body as an extended thing to a specific distance from where I was a moment ago. I move my body by borrowing other bodies, the ground, the objects that stand between me and the bookshelves, and the air that I penetrate with my body. This displacement is dependent on place, on extended space (*res extensa*). My body deploys itself and is localized in space, that is, in this dense and "flexibly solid" substance which makes my movements possible. Where, then, is the subject? Is it in my body, where it enacts its movements, or does it dwell elsewhere? This elsewhere can itself only be a place, and thus extension. What is the gap separating the subject from the object? Do they have the same substance? Wherein lies their difference? How can a *thinking* substance (*res cogitans*) be related to a radically different substance insofar as it is *extended* in space? This way of conceiving the relationship between man and the world and objects will probably not enable us to resolve the relationship between them, for it rests on a pre-established difference that forbids any ordinary commerce between them.

When I am about to go to the other end of the room, I am *already* there even before I put my decision into effect. Without this "already-being", no act would be possible. But the adverb "already" cannot be reduced to a "before", following which there would be an "after". As soon as I think about this displacement, I am immediately in the place I am thinking about. When I think of the Pantheon – from here, from my workplace, from where the monument is not visible – I am already there in a sense, for my mind gives presence to the Pantheon thing in its distance. This proximity of my being with regard to the thing is the condition of all distance-taking. Without it, no distancing would be possible. A distant thing could not be conceived of in terms of its distance if it were not already given to me. The subject is not an enclosed interiority that then opens itself to the thing before taking shelter again in his closedness. He is *always-already* there. He is not a here that is subsequently there. His being resides in the already-there, in his *Da-sein* (being there), Heidegger says. With regard to our example, there is an objection: "It was Alain who pointed out," Lacan (1975b) writes, "that no one counts the number of columns in his mental image of the Pantheon. To which I would have liked to have answered him – except the architect of the Pantheon" (p. 143). Lacan's remark echoes what in phenomenology is called the *intentional character* of consciousness. Hence the principle that only the architect of the Pantheon could be concerned by the number of columns.

We will see further on that these affirmations are borne out by the physiology of movement. Space, therefore, is not simply a matter of extension. It is closely bound up with a kind of temporality that calls for further elaboration. This requires a very different approach from that of the natural sciences which content themselves with the truism of the linearity of time.

Time as simultaneity

The above example clearly shows that my path from my desk to my library does not describe a set of points placed one after the other. As space, it rests on a mode of apprehending non-linear time. I do not perceive point A first, and then point B. My perception seems to distend itself continuously from A to B. It is what Saint Augustine (1960, *Confessions*, Book XI, 13–28) calls the *distension of the soul*. Along my path towards point B, my mind *tends* as much towards the point of departure A as it aims for the point of arrival B, and it does this from the present now. These three moments of present, past and future are not subject to any time difference between them in such a way that they could form a linear succession. Husserl qualifies them as *momentary phases of consciousness* linking the past to the future from the now of the present.

Consciousness of time, according to Saint Augustine, is dependent on memory. Husserl, too, adhered to this conception in his inquiries prior to 1909. But, according to the later Husserl, this approach to time fails to distinguish between memory and the perception of time. *Time is clearly different from the memory of time.* Memory puts the temporal object in the background by relating to it *actively*, whereas the perception of time retains the past moment of the object *passively and intuitively*. Henceforth, Husserl abandoned the Augustinian thesis and substituted the term *retention* for that of memory. The latter concerns a completed process in time, whereas retention takes account of what is still present and unfolding. Likewise, after 1909, Husserl (1964, p. 51) qualified consciousness of the future as *protention*. When we listen to what our interlocutor is saying to us, we do not memorize what he has just said in order to understand what follows. Rather, we *retain* the past elements of his discourse at the same time as we "pro-tain" (i.e. anticipate) what he is about to say. We are thus in a position to grasp in *praesentia* the flow of what is unfolding. The three moments of the past, the present, and what is to come, are apprehended *simultaneously*.

Aristotle and the question of time

"No attempt to get behind the riddle of time," Heidegger (1982) writes, "can permit itself to dispense with coming to grips with Aristotle" (p. 232). Aristotle was the first philosopher in the history of human thought to question time decisively and to give it the impulsion that it has enjoyed since Antiquity.

Aristotle (1936, *Physics* (Δ 10, 217 b 29) begins by questioning the concept according to which time is identifiable with movement or motion. Encompassing all existents,[1] the celestial sphere is the moving force par excellence by virtue of which the objects of the world enter into movement. Aristotle does not stop at this simplistic explanation. Concerning movement (κίνησις), he asserts that time is "something like a movement". But movement is always in the moved, that is to say, in the moving force. Admittedly, time is not itself movement, but it is not without movement either. Time is thus "something of movement" (κινήσεώς τι). What, then, is the relation between time and movement? Aristotle replies by asserting that time is "number" (άρθμός). "Time is the number of movement," he writes, "in respect of the before and the after." Movement is measured by time; but, equally, time is measured by movement. Consequently, time is a numbered number which, as something of movement, places itself within the horizon of the anterior and the posterior. This intrinsic relation between time and movement sullies the question of the temporality of ambiguity by making it dependent on spatiality. According to Heidegger this ambiguity is inherent to the Aristotelian term number (άρθμός) which signifies in turn adjustment, number and quantity, but also the length of a path, the duration of a period of time, counting and numeration (Heidegger, 1982, p. 235).

What is movement? If one were to define it as a change of place, there would be a risk of confusing change and translation. Not all movement involves a change of place. Let us return to time as number. According to Heidegger's interpretation, time, as number of movement, is numbered in the movement. Time is not, however, in the movement of the hand of a watch, or in its figures and numbers. At nightfall, time is not in the horizon that I am looking at. Nor is it in the watch that I am consulting. Where is it then? Aristotle places it within the horizon of the anterior and posterior. This is what he comes back to constantly and it is here that he never loses sight of the constitution of time. Consequently, the horizon means that it is there that we unfailingly find time. It is on the horizon, Heiddegger says, that time comes to meet us. Is it a tautology, he asks, to state that time is the before and the after, i.e. time is time? This before and this after, are they indicators of place? Is the numbered movement to which Aristotle refers to be understood as a change of place? Answer: no, alteration is also movement without necessarily being translation. An alteration occurs at the same point and within one and the same thing. In alteration, however, there is also an anterior state and a posterior state. Heidegger explains: "We shall call this structure of motion its dimension, taking the concept of dimension in a completely formal sense, in which spatial character is not essential" (1982, p. 242). He traces the dimension back to its origin, namely, tension (*Dehnung*). He then adds that extension (*Ausdehnung*) in the spatial sense of the term is merely a modification of the dimension involved. "In the case of the determination of ἐκ τινος ε'ίς τι," he writes, "we should rid ourselves completely of the spatial idea, something that Aristotle did, too. A completely formal sense of stretching out (*Erstreckung*) is intended in 'from something to something'" (1982, p. 242). Here, Heidegger is concerned that the extension in question should not be reduced to the Cartesian *res extensa* (extended thing).

It goes without saying that the flow of time, its continuum, is an integral part of its structure. "When we experience motion in a moving thing," Heidegger explains, "we necessarily experience along with it συνεχές, continuity, and in this continuity l'ἐκ τινος ε'ίς, dimension in the original sense, stretching out (extension)" (1982, pp. 242–243). Extension and continuity are therefore already involved in movement. The latter follows (escorts) dimension, that is to say, extension. This takes us back to Augustinian distension and the intentionality of Husserl in the retention of the past and the protention of what is yet-to-come.

Following movement comes the numbering of the nows, that is "the before, the no-longer and the not-yet". According to Heidegger, for Aristotle, the nows are co-perceived and consequently co-numbered within movement. In this respect, they themselves are not only numbered, but also numbering, for they number the places traversed by movement.

Time, Aristotle says, is the numbering number that is itself numbered in the horizon of the before and the after that is inherent in it. This means that it contains within it distance in relation to the now, which is the limit of what is past and what is yet-to-come. The now that we number is different each time. While being always the same, it maintains itself constantly outside itself. Thus it is extension par excellence. "The ever different nows," Aristotle says, "are, as *different*, nevertheless always exactly *the same*, namely, now" (cited by Heidegger 1982, p. 247).

In repetition, too, there is the identical, namely, the succession of the identical to itself. But our natural attitude only envisages time from an unexamined point of view, that is, as an *unchanged* succession of the same. True, repetition constitutes the distinctive characteristic of the living. Hence the predominant idea of equating it with the machine. But repetition is the renewed resumption of the same, without the latter being reduced to the identical. It is characterized, in fact, by the principle of mismatching (*écart*).

The constitutive relation between time and number continues to raise questions for us. Is repetition the *nagging* succession of numbers giving rise to the flow of time? What is the nature of such a succession? Is repetition always the unchanged return of the same, namely, the recommencement of the now? Is it really the case that there is no change in this succession of nows? If that is so, how could there be any change?

Lacan (1961–1962) deliberately chose the term *single trait* (*trait unaire*) to render the Freudian term *einziger Zug,* a term Freud uses to describe one of the principal characteristics of the mechanism of identification. According to Freud, in identification, the subject does not repeat every feature of the person with whom he identifies. He takes over a characteristic trait, an *einziger Zug*, a single trait. The Freudian question is one of knowing how this single trait, that is, this particular "one" makes its way to the other person and constitutes itself as *identical* to the identification. The question is explored in a different manner by Lacan.

Taking up the question of the mathematician Gottlob Frege concerning the passage of zero to one, Lacan (1961–1962) refers to the traits engraved on an animal bone that he had contemplated at the National Archaeological Museum of St Germain en Laye (session of 28 February, 1962). The traits are identical. But Lacan argues that this is not so. He explains how prehistoric cavemen, the supposed authors of the markings, were doing nothing else in their act of counting than establishing difference at the heart of the second trait in relation to the first. If the counting concerned, for instance, the number of their wild animals, the figure two, succeeding the one, differentiated it fundamentally from the first. By marking the second trait, the prehistoric cavemen acceded to one, then two, and so on. Difference then emerged at the heart of the identical. The single trait is consequently what differentiates the one by virtue of its identity with the other. That is to say, the constitution of identity will be deferred again and again. Difference is established retroactively as "differ*a*nce" (Derrida, 1967). I will be discussing the temporality inherent in the latter throughout this book.

Common time and the prior access to time

We are now going to apply the question of the engraved traits, discussed above, to the now of the temporal flow. These nows follow each other, then, but without resembling each other. By keeping time present, the no-longer now and the not-yet now are both constituted. This unity of time that is the now is both numbered and numbering. This is how movement occurs. The movement of the hand on the watch illustrates this point.

But, Heidegger objects, time as movement numbering the number merely describes, in fact, our *condition of access* to time. It therefore only expresses the succession of now-points according to a traditional approach, that is, the *common* conception of time. In conformity with the latter, time is not given to us; rather, it unfolds by virtue of the now-points that follow one after the other. This conception rests on the *forgetting* of time insofar as it is given to us *in advance.* How, indeed, could we have *access* to it if it was not at our disposal in one way or another? It is this same prior donation of time that enables us to find it on the clock on which the second hand, precisely, marks the now-points. Does that mean that time is a prerequisite for our existence?

We will have to see what happens in the interval from one now-point to another or in the distance that differentiates the second trait from the first. Of course, they repeat each other and establish themselves in an identical register. Wherein lies their difference then? How could such a distinction emerge from a tautological relation? What happens when one passes to the other? The answer does not lie in the instant of their transition. Rather, it is to be sought in the instant before, that is, in the identical that is the first trait, the first now. For it should contain *within itself* something that allows us to pass to the following moment. What is it, then?

The one of the first moment should prompt us *intrinsically* to pass over to the other. It lies in the first instant, in a single trait, an incitation, an invitation, an irresistible impulse leading us to envisage the second, and so on. This "thrust" can only be in keeping with our nature in such a way that we cannot proceed in any other way than by welcoming the *following* moment. It is in this same act of welcome that the *preceding* moment is constituted. Time, Aristotle said, unfolds within the horizon of the anterior and the posterior. We will see that this temporal condition is also attested by neurophysiology.[2]

If we look closely, in our dealings with the now there is always a particular mode of expecting, namely, expecting a sequel inviting us to go towards the *encounter* of what is to come. It is not an expectative waiting. Something *puts* us in a state of waiting-for. "Such a being-expectant (*Gewärtigen*), an expecting, expresses itself by means of the then" (Heidegger, 1982, p. 259).

Time as expecting is what is given to us beforehand so that we are capable of watching out for, of looking for, the imprint (*Prägung*) of a trace that is yet-to-come. This pre-givenness of the time of expecting is thus in keeping with so-called objective time, expecting that depends on an *entente* with reality. Whether it is marked by fear or joy, time accompanies us in everything and in all circumstances. It is neither contingent nor fortuitous. Everything is inscribed in it and everything takes shape and consistency in it. What, then, is this expecting, constitutive of time?

The mismatching (*écart*) that I have pointed up as being the distinctive characteristic of living things (*du vivant*) is simply a consequence of this expecting. But let me add right away that the latter does not have a teleological or eschatological character. It is a fundamental aptitude of man in particular and of living things in general for *waiting-for (s'attendre-à)*. Whether it is called instinct or drive, it is always a *prompting* that comes more from the posterior than from the anterior, more from the after than from the before.

The common understanding of time leads us into error, even in its scientific consequences. An animal does not respond to a stimulus. It receives it on the basis of a prior disposition that prompts it constantly to *wait-for* . . . In other words, it merely responds to the call of what follows, of what comes after. We have become accustomed to referring to this precondition by terms such as motivation, drive, vital force or instinct, without thinking any further about it. In so doing, we attribute animals with a vacuity that is prey to external stimuli. In the instinct as in the drive, there is neither pure passiveness nor pure activity. However frightened an animal may be when faced with danger, it either puts itself *on its guard* or *envisages* flight. Its reaction to the aggressor is in keeping with the danger. It is not that aggression comes first and is then followed by reaction. The latter would never have existed if the animal had not been disposed to it in advance. In what it is expecting, there is an "agreement", a concordance with what we commonly call stimulus. An animal is thus *in tune* or *in "affordance"* with that towards which it strives or in the face of which it withdraws.

The term *hereditary coordination* designates the almost invariable sequence of phenomena culminating in the accomplishment of an instinctive act. The triggering of each of its "links" occurs not only in accordance with the one that precedes it, but above all in accordance with the one towards which it strives. In the absence of external stimuli, the lure can trigger the same succession of acts. It functions as a real triggering stimulus to the exact extent that it offers what the animal is expecting.

Likewise, what an animal learns through conditioning is what it learns to expect. So as soon as mention is made of something we would like to taste, our "mouths begin to water" just as if we were expecting to taste it.

In order for a stimulus to reach an animal, the latter must already be open to it. It is the following stimulus that calls the one before. This is what Lacan (1961–1962, session of 24 January, 1962) calls, not without a certain elegance, an "error of counting", namely, taking the zero for a unity in order to gain access to the one. Without such an error, there would have been no access to the number or the countable. By taking itself for a unity, the zero gets ahead of itself as if it were already the one that it would like to bring into being. The precedence of the response over the stimulus is borne out by studies in ethology.

Time and ethology

By placing the notion of mismatching at the heart of the question of the living, I have been led to question temporality insofar as it stands for the state of *waiting-for* (*attente*). It would be legitimate to see if the latter also applies to the animal soul. This leads me now to present an overview of ethology as a comparative study of animal behaviour and human conduct.

Almost one hundred years ago, Nicolaas Tinbergen, one of the pioneers of ethology, opened up the question concerning animal instinct with a view to assigning it the place it deserved. Not long after, a disagreement emerged between continental ethologists and those from North America, for the large part behaviourists. For the former, it seemed absurd to accept that an animal was merely an ensemble of stimuli/responses; whereas for the second, it was sufficient for the new science to focus on the study of an animal's visible behaviour, ignoring all unverifiable concepts such as perception, memory, emotion, or reasoning.

Tinbergen nonetheless taught us that instinct was not a rigid and inflexible entity reducing an animal to an automaton. Instinct is governed by time, not only because the organism of an animal is subject to well-defined temporal cycles, but because it contains time and temporality within itself. One of the major discoveries in ethology was the phenomenon known as imprinting (*Prägung*). It is by virtue of such a disposition prior to all experience that an animal makes its choice of an object of *attachment* as soon as it is born, provided that the object conforms to what it is *expecting*. In imprinting, the subject is, so to speak, the stamp or impression (*Prägung*) of its object. This openness precedes the arrival of the object. In this respect the latter is an empty entity. We saw this earlier in the schema of the partial drive. It is precisely the empty character of the object that causes the insistent repetition of its movement to-and-fro. But, however empty the object may be, it must be in tune with the orifice (erogenous zones, Freud says) that receives it and which turns around it. This is how the breast or the nipple is in tune with the orifice of the mouth. It is this matching that Gibson, the founder of ecological psychology, calls "affordance". The branches of a tree are made for birds as is a stone for the serpent. Affordance signifies how the thing "lends itself" for our use. Consequently, affordance is the disposition of the thing to let itself be in tune with the deeds and gestures of the subject who will make use of it. It is by putting itself at our disposal that it can be self-determining. The object's affordance does not mean that the latter complies with our usage. On the contrary, it is the subject that submits himself to it by vanishing behind the object. In the act of affordance, the thing frees itself. And it is by freeing itself that it frees everything that it has at its disposal.

The birth of ethology during the twentieth century is the logical extension of the episteme of modern times in which man occupies the central place. And so the young science inevitably found itself caught up in the debate opposing the vitalists and the mechanists. Owing to its conception of reducing living things to the system of stimuli and responses, the American behaviourism of John Watson immediately claimed allegiance to the mechanist current. He resisted all recourse to a vital and supernatural force for explaining animal behaviour. However legitimate their struggle against vitalism was, in their radical approach the behaviourists neglected any form of endogenous force in the living being. The Pavlovian model of the reflex arc served as a basis for refuting any other activity capable of explaining behaviour.

The European pioneers of ethology, all vitalist and mechanist tendencies included, found themselves faced with the behaviourist truism whose practice boiled down to training animals by conditioning, a practice that was not exempt from violence (see Renck and Servais, 2002, particularly the first part). Just as the American ethologists were working essentially in the laboratories, their European colleagues spent most of their time in the company of animals. Thus European ethology adopted from the outset a phenomenological approach to the study of animals. It was this common trait that brought vitalists and mechanists closer together, in spite of their divergence of principle. Vitalists such as Jakob Von Uexküll and Frederik Buytendijk never lost sight of the importance of scientific rigour. Oskar and Katharina Heinroth, as well as Wallace, Craig, and Nikolaas Tinbergen – the pioneers of European ethology – never neglected the subjective universe of animals in their investigations. But the chief merit lies with the father of modern ethology, Konrad Lorenz, for his effort in combatting behaviourism without losing sight of the scientificity in his investigations where subjectivity and rigour tend to join forces.

The first idea ethology wrestled with was that which concerns the immutability of instinct. F. Buytendijk, one of the pioneers of animal psychology in Holland, belonging to the Husserlian current of ethology, had already affirmed in the 1920s that "many instincts are absent or imperfectly present in young animals and are developed spontaneously or through experience" (Buytendijk, 1920, p. 118, translated for this volume). Refuting the opposition established by Bergson between instinct and intelligence, Buytendijk said that it was preferable to define instinct in the broadest sense as the psychic basis of a complex of actions that an animal executes by nature (innately), under the special conditions of time and place, actions stemming from the external stimuli and internal influences of the organism.

(1920, p. 109)

Buytendijk's definition took into account several basic notions of ethology. First of all, it excluded any direct relationship between consciousness and instinct, defining the latter as "a psychic base of a complex of actions". Next, it made instinctive development dependent on external conditions, namely, time and place. This dependence is based on one of the essential notions of ethology, *taxis*, which is the translation of vegetable tropism in the domain of animals. Claude Bernard had already established the fundamental irritability of the protoplasm as the cause of movement and orientation in living things, distinct from other aspects of cell metabolism. Research in botany, initially, and then in ethology, succeeded in determining to a large extent the mechanisms of tropism and taxis. Gravity, light, humidity, and differences of potential are among the main causes of the orientation of living organisms in space. Depending on the reaction of the latter and the direction of their irritability when faced with these external elements, one speaks of positive or negative tropism.

Taxis refers to an animal's deep affinity with its surrounding environment. Thanks to this kind of affinity, instinct acquires a certain plasticity. This was demonstrated by Charles O. Whitman, another pioneer in ethology, in his detailed research on pigeons at the beginning of the past century. According to him, instinct conserves, in spite of its invariability in the species, a certain aptitude for change due to experience.

Wallace Craig, Charles Whitman's pupil, followed up the research carried out by his teacher. He succeeded in laying the foundations for the understanding of instinct as an expression of the affinity that exists between an animal and its environment. Opposing the dominant behaviourist outlook, Craig refused to apprehend instinct as a sequence of conditioned reflexes. He divided instinctive behaviour into two constitutive phases: appetitive behaviour and consummatory act. Initially, an animal engages in a series of actions that lead it to encounter the elements and situations liable to facilitate the execution of its instinctive behaviour (appetitive behaviour). The search for stimuli necessary for awakening its instinct is part of such a phase, which culminates in the *awaited* consummatory act, which is both innate and perfectible. Wallace Craig completed this succession of instinctive acts by two other phases: the phase of overabundance engendering the aversion of the animal towards its stimulus and the state of rest that follows it (see Craig, 1918). Appetitive behaviour signifies that the animal is *searching for* stimuli capable of provoking its consummatory act. In other words, during this phase the stimulus is actively awaited by the animal. "Quite generally," Lorenz (1978) writes, "releasing mechanisms, whether innate or acquired, are programmed so as to set off adequate behaviour at the teleonomically correct time. In this respect innate releasing mechanisms (IRMs) can be regarded as mechanisms exploiting instant information" (p. 243). As for the aversion of the animal, it goes without saying that at this stage the stimulus is no longer awaited, that is to say, the animal is no longer in an attitude of expectancy.

This being so, the overlapping cycles that Craig saw behind an animal's quest concerning its sexual partner or its food are the expression of a particular relationship to time. The latter seems to be constitutive of what Lorenz calls the *motor schemas* of an animal in its instinctive behaviour. The most striking example is that of so-called violinist (fiddler) crabs whose activities are regulated by lunar time, at intervals of 12.4 hours, that is, in accordance with tidal rhythms (see Palmer, 1988). Research studies seem to have established that this relationship to measurable time is an endogenous process that is nonetheless regulated by environmental stimuli, that is to say, by cosmic time. There are many such examples showing the rhythmic existence of time within the activities of living beings. The most familiar case is that of menstruations in women. Some research has established that the organ responsible for such *circadian rhythms* is the pineal gland in mammals. Other studies put the emphasis on the genes responsible for these internal clocks (see Piccin et al., 2000).

Although it depends on time, the biological clock does not place animals in a relationship to temporality. In an animal that is endowed with it, time functions as an automatism. *It can, however, participate in the project of the living.* This is the case, for instance, of menstruation in human beings. The same is true of the circadian rhythm that regulates the periodicity of sleep. Here as elsewhere external reality can suffer from the subjective provided that it participates in the *project* of the living, which is always ahead of itself. This mode of being cannot be reduced solely to the quest for a goal. However determined its *project* is, the living remains open to the unexpected, to the unforeseen. We owe this vital law to the genius of Darwin. Moreover, it was by means of such an attitude that Craig defined the appetitive phase as instinct; for it is the project designed to suit what manifests itself as stimulus that is capable of triggering an act in waiting.

The essence of expecting lies in its relation to the unforeseen, to the unexpected. The encounter of the tick with the warm blood of mammals is in principle contingent. But as soon as it happens, it triggers a sequence of events which gives rise in the most determined way to the affordance between an animal and its environment. The consummatory act is triggered following the period of appetite. It puts the animal's behaviour in tune with the external triggering stimuli. Lorenz writes,

> The feeding behaviour of the Peregrine Falcon is predominantly based upon innate motor co-ordination patterns. Appetitive behaviour is restricted to the search for the requisite stimulus situation, operating according to the principle of *trial and error*, and the exquisitely specialized instinctive behaviour patterns of prey-catching typical of this bird species are elicited when the appropriate situation is reached.
>
> (1970, p. 278)

The trial and error of which Lorenz speaks is required by the very readiness of the animal to go out towards the world, and this is due to its being ahead of itself.

Insofar as the unforeseen leaves its imprint on expecting, it guarantees the inherent gaping hole in the latter. It guarantees the gap involved in expecting as a distinctive trait of the living. The unforeseen is a form of chance and curiously it seems to determine the instinctive act. Chance should thus be understood in the sense of what Aristotle calls *tuché* (τύχη) and *autómaton* (αυτόματον). Let me illustrate the Aristotelian concept with an example. When a road accident occurs, the injured see an ambulance arriving even though they have not called for it. There is *autómaton* (chance) for the ambulance and *tuché* (fortune) for the injured. Chance, in this sense, is the *interplay* between the necessity recognized after the event (*après-coup*) and the contingency of the event. In other words, these two factors constitute the condition of chance insofar as it culminates in instinctive determinism. Instinct is the meeting (*autómaton*) between the spontaneous (endogenous) processes of the animal and its external triggers. It is what moves by itself (γένεσις αυτοματόσ) thanks to its encounter with its surrounding stimuli. Thus expecting, in the strict sense of the term, is the openness of the living being insofar as it is dependent on chance and the unforeseen. It is not teleological, but takes on the appearance retrospectively of a goal programmed in advance. I will come back to the question of chance later on in my study of dreams in order to explore it more fully.

The phenomenon of *insight* in animals is further evidence of this openness that we call waiting-for or being-expectant (*attente*). Faced with a problematic situation, an animal ceases all activity aimed at solving it, then throws itself into a sudden discovery in order to resolve the situation. Lorenz emphasizes the relation between insight and curiosity. According to him, curiosity is the highest level of activity of living things. The search for the unknown out of curiosity highlights, in fact, a specific mode of *expecting* in which the aim does not have the serious character that we recognize in teleological activities. According to Lorenz (1978), "This phenomenon is equally characteristic of both exploratory behavior and play, and thus it cannot be used to separate them by definition" (p. 327). Playfulness is, in fact, the condition of all curiosity. But as Freud taught us, play is not the contrary of the serious. It is in such a playful relation that the living being gains access to questioning, which constitutes his highest level of openness to himself and the world. We shall see that it is precisely in its quality as drive that the dream constitutes itself as playful activity.

Freud and Lorenz, drive or instinct

The French term *pulsion* comes from the Latin *pulsio* (the act of pushing and repelling). According to the *Dictionnaire historique de la langue française*, the term

> acquired a specialized sense in the eighteenth century in physics to designate the propagation of movement in a fluid and elastic milieu, a usage attested in 1738 by Voltaire, probably modelled on the English pulse, representing (c. 1330) the Latin *pulsus* which had been ascribed a particular meaning in physics by Newton in 1673.
>
> (Rey, 1992)

It was probably this latter meaning that governed the usage, in the psychology of the nineteenth century, of its German equivalent *Trieb*. It was in his *Three Essays on Sexuality* (1905) that Freud used it for the first time as a psychoanalytic term that was to undergo numerous elaborations. Its English equivalent "drive" appeared in the *Encyclopaedia Britannica* for the first time in 1918 in an entry by the American psychologist Robert S. Woodworth. But in a more general way, the term "drive" imposed itself at the beginning of the twentieth century under the influence of the behaviourist movement.

It was to a large extent over the question of the difference between drive and instinct that the European ethologists distinguished themselves, under the aegis of Konrad Lorenz, from their American colleagues. For the latter, the drive encompassed the ensemble of innate behaviours, provided that they were considered at the functional level. "In America," Lorenz wrote in 1937,

> where the use of the word instinct has recently fallen from fashion, the terms "first-order drives" and 'second-order drives' are used in the same, or very similar, sense as the terms 'instinct' and "motor mechanism" are used by McDougall. McDougall and a number of more recent authors regard the relationship between the two, which is assumed to be based on the employment of the innate motor co-ordination pattern as a means to an end in the service of a governing instinct directed towards a particular goal, as proof of genuine purposivity in the first-order instinct.
>
> (Lorenz, 1970, pp. 292–293)

European ethology distinguished itself henceforth by putting the emphasis on the instinct in the most naturalist sense of the term.[3] To this end, it focused on the study of innate behaviours as a distinctive trait of animals, just as zoology focused on animal morphology; whereas the American current was more interested in the question of learning and of the functional behaviour of animals. The chief reproach of European ethologists with regard to their North American colleagues consisted in saying that they contented themselves with studying animal behaviour in their laboratories rather on the ecological terrain where an animal manifests its primary affinity with nature. Going against the primacy accorded to learning, Lorenz put the emphasis on instinctive behaviour as such. "I consider it as an important characteristic of the instinctive behaviour pattern," he wrote,

> that it can achieve results which are beyond the intelligent capacities of the animal species concerned. For this reason alone, it would seem to be impossible for an animal to improve its own instinctive behaviour patterns through learning or insight. In practice, we are unable to decide absolutely whether an instinctive behaviour pattern is fundamentally unmodifiable by learning or insight. . . . *The solutions to the problems set in the environment of the animal achieved by instinctive behaviour patterns are always far in excess of the intelligent capacities of the species.*
>
> (Lorenz, 1970, p. 290, my emphasis)

Can as much be said on the subject of the drive in man? Do unconscious drive impulses involve such a challenge in the creatures endowed with language that we are? Some elements of an answer to this question may be found in my discussion concerning the oneiric impulse. Indeed, it is in dreams, where consciousness sinks into obscurity, that the logic of the unconscious reveals itself with all its force.

In a comparative study between instinct and drive, it is important to focus on the constituent elements of each of them. We have seen that Craig divided the instinctive act into three parts: the appetitive phase, the consummatory act, and satiety. What about the drive in Freudian theory?

Would it be legitimate to approximate the appetitive phase to the period prior to the so-called latency period on the one hand and to adolescence on the other? Bringing these two periods together under the same banner might seem incongruous. But it would be justified in terms of psychoanalytic theory. It is here that the very principle of the drive in the Freudian sense of the term comes into play. According to phallic primacy, a cornerstone of psychoanalysis, the signifier of the phallus is the main constituent in the psychic structure of the individual, whether man or woman. This primacy being established, the infant constitutes himself, as early as the oral stage of the drive, that is, from birth, in relation to the phallic lack of the mother. In other words, he is supposed to make up for the maternal lack imaginarily. But the intervention of the law (represented by the father) puts a halt to this incestuous relationship by forbidding the child to enjoy his mother's body, the prohibition of incest being a universal law that plays a foundational role in human societies. Having passed through the Oedipal stage proper (around the age of 5), the child will be able to find his subjective place with regard to his instinctual drive aspirations. This allows him to take leave momentarily of his instinctual drive conflicts and to enter the latency period, between the ages of six or seven, up to the onset of puberty. During this period, drive energy is invested in educational, social and cultural learning, leaving the earlier psychic conflicts in the background. Adolescence is the reactualization of these same Oedipal conflicts in a critical sense, favouring the young person's initiation into the adult world. It is here that the phase of appetite would have its full meaning, culminating eventually in the consummatory act. But, according to Freudian theory, this is not what happens.

The drive, according to Freud, does not seek the consummatory act except in the form of doing away with the state of tension. The *principle of inertia* means that it has the tendency to return to the level zero of excitation. Pleasure, according to Freud, begins where unpleasure ends. Furthermore, repression is the necessary mechanism by means of which the drive constitutes itself as such. Sublimation is another solution, akin to what Lorenz says on the subject of play and curiosity. The vicissitudes of the drive are thus multiple and dynamic. Sometimes it undergoes repression, sometimes it is transformed into a socially useful product (sublimation), and sometimes even into its contrary. In comparison with instinct, the lability of the drive is all the greater in that it espouses the formation of signifiers. From his conception, the human child is immersed in language, which marks the drive with the stamp of desire.

This tendency to return to the point zero is nonetheless experienced by the human subject as a danger. This is what is suggested by the term *aphanisis,* the fear of seeing one's desire disappear. A struggle begins aimed at stopping regression and the disappearance of desire. At this stage, the subject appeals to phantasy, which is the principal support of desire. A man experiences aphanisis as a real castration; in a woman, this is expressed by the fear of not being loved.

What Craig calls aversion may be likened to aphanisis. But while the first is accepted by an animal, which takes it as the natural term to the process of its instinct, the second initiates a relationship marked by struggle and conflict in man.

If the consummatory act exists, it is dependent on this aspect of the imaginary that is phantasy. The latter is accomplished as a matter of priority, that is to say, in all haste, even if it respects the instinctual sequence of the act. Haste is, in fact, as we shall see later on, the temporality of phantasy. It is a singular form of waiting inasmuch as "it cannot wait". This is the very meaning of the drive with the precipitation that we recognize in it.

One of the common characteristics between the human drive and animal instinct is the imaginary dimension. Ethologists speak about it under the term lure. That the Barbary duck, *caïrina moschata,* defends the little mallard duck because it emits almost the same cry of alarm as its own offspring, does not differ, in its essence, and on the imaginary level, from the excitement of human beings in front of pornography. In both cases, it is the role of the lure that determines the act of the drive. Precipitation, that is to say, misappraisal, guarantees the efficiency of the lure that is so essential for warding off the aphanisis that fatally plagues every act of the drive. It is such a phenomenon that Craig calls the phase of aversion preceding the state of rest.

Generally speaking, the animal world is governed throughout by the reign of the imaginary. This is borne out by fishing instruments that are equipped with bait for fish. "Geese can react," Lorenz (1978) writes,

> to a leaf wafted along by a slight breeze as if it were a slowly gliding eagle. I have known incubating turkey hens to roll smooth pebbles or − in one case − a tin cigarette box into their nests because the innate releasing mechanism (IRM) of egg rolling responds to any object that is hard, smooth, and devoid of projections.
>
> (p. 81)

The IRM cannot be apprehended without taking into account the animal's aptitude, however rudimentary it may be, for imaginary functioning.

The goal that is always in question when speaking of instinct presents itself quite differently in relation to the drive. In the closed circuit of the drive which I referred to earlier, the drive begins its trajectory from its somatic source, called an *erogenous zone,* before reaching its final point which Lacan designates with the English term "goal". The loop or the closed circuit of the drive engenders a movement back and forth, for each time the aim of satisfaction (the goal) is not attained, but missed; hence its ever-renewed departure. "Let us concentrate," Lacan (1973a) writes, on this term *but*, and on the two meanings it may present. In order to differentiate them, I have chosen to notate them here in a language in which they are particularly expressive, English. When you entrust someone with the mission, the aim is not what he brings back, but the itinerary he must take. The aim is the way taken. The French word *but* may be translated by another word in English, goal. In archery the goal is not the *but* either, it is not the bird you shoot, it is having scored a hit and thereby attained your *but*.

(p. 179)

This closed circuit designates what Freud calls the partial drive in order to differentiate it from the genital drive on which the sexual and subjective identity of the human being depends. The montage of the partial drive seems to be in conformity with what ethologists describe at the level of the consummatory act of the instinct. They insist on the infallibly repeated and ordered sequence of this act, which seems in keeping with the traced trajectory, the aim, Lacan says, of the drive loop.

The term "goal" seems to explain astuciously the kind of finality involved in the instinctive act. The embarrassing question of finality debated between vitalists and mechanists thus finds a partial response. Lorenz (1970) gives, among others, the following example:

It can easily be shown that even in the raven the concealment response is an adequate goal in itself since this response will be performed to excess in a non-adaptive and non-functional manner under the requisite conditions of captivity, just as by the jackdaw.

(p. 288)

The idea of goal, moreover, as a specific rather than a teleological genre, is perhaps not far from what is generally called teleonomical (Lorenz, 1978, p. 34), designating the end to be attained in the instinctive act.

In the closed circuit of the partial drive, there is another constitutive element to be determined. The loop never ceases to close itself again around what engenders its movement, namely, the hallucinated object. The latter is the paradigm of what Freud calls the lost object, which must be refound over and over again. We know that this object is not lost, even if it has to be refound. This sensation of reunion in the drive is named most judiciously by Freud as *Drang*. It signifies the force, the urgent pressure, what thrusts and seeks to attain its aim as a matter of priority, what can no longer wait, that is to say, *waiting par excellence*. The heart of the meaning of *Drang* is understood by Lorenz, who writes:

The emergence of restlessness in the animal – the *impetus* for directed or undirected seeking for a specific stimulus situation, in which the innate releasing schema of the requisite response is first brought to elicitation – is exactly what I would describe with the word *drive*. (I am completely aware that this concept of drive is even less customary than the concept of the instinctive behaviour pattern which I am employing.)

(Lorenz, 1970, p. 310)

What is an animal striving for (*anstreben*) in its instinctive behaviour pattern, Lorenz asks himself? "For a stimulus situation," he replies, "stemming from pleasurable sensations . . . of an earlier experience accompanying the performance of the instinctive behaviour pattern" (ibid). The lost object, the reunion with which is always *awaited*, may thus be considered as the constituent essence of every instinctive behaviour pattern.

The closed circuit of the partial drive, with its aim, its goal, and its empty hallucinatory object has its equivalent, it may be argued, in what ethologists call *vacuum activity*. Lorenz gives the example of a starling in captivity which performed imaginarily all the behaviours connected under normal conditions of liberty with the hunting of real insects. The "pseudo-pregnancy" of domesticated bitches also attests to vacuum activity which originates in the innate drive of the animal. Vacuum drive activity occurs in situations of captivity and domestication, that is, in the absence of real releasing stimuli. This absence is responsible for the lowering of the threshold of instinctive releasing. Empty activity, therefore, is linked to a deficiency. Is it the same kind of deficiency that can be observed in the releasing of so-called stereotypic auto-erotic activities in man? We know that the peak of such a deficiency is found in a child suffering from hospitalism, where the human infant is lacking the maternal Other, a situation that is perhaps close to that of captivity for an animal. Such a deficiency can prove lethal, that is to say, it can switch the life drive into a death drive. We do not know if this is also true of animals. The question is one of knowing whether, in the absence of its object, a living being does not turn against itself and let itself waste away from despair in relation to what causes its expectation. A comparative study between ethology and psychoanalysis could broach these kinds of questions.

Lorenz, the father of ethology, and Freud, the father of psychoanalysis, had many things in common. Both of them had benefitted from the same university education and had studied medecine. Both considered their new discipline as a legacy of Darwin. Neither of them, however, fell into the trap of mechanistic thinking. They also saved themselves from vitalism while remaining faithful to the subjective dimension without which their respective domains of research would have lost their raison d'être. Both were able to keep their thought free from the all-powerful American vision of knowledge which began to extend its supremacy on a worldwide scale. Their investigations proved to be of high scientific value, while remaining accessible to the public at large. In their respective studies, they were able to transcend the age-old idea that opposed instinct and intelligence.

Psychoanalysis and ethology, a relation of *intension*

Freud's contribution concerning the drive drastically changed our view of man in quite another way. The drive determines "the initial vital funds" of the human subject. This formidable life force (*élan vital*) is paradoxically confined to the destructive force of the death drive. This antagonism stems from the constant nature of conflict that Freud found to be at work everywhere in man. The drive, struggling with the dispute inherent to its unconscious forces, is characterized by slippage, and in such a constant way that one cannot distinguish drive and derivation; the same is true of the signifying chain, constitutive of language, the permanent reference to which knows neither rest or arrest.

Freud's concern was to establish the quite natural link between body and mind, a link that had lost its foundations and its pertinence since Descartes. The drive is this thrust (*Drang*), that compelling force that surges forth spontaneously and which hastens to accomplish itself. In order to realize its purpose, it espouses the representation (the signifier) that will be its representative (*Vorstellungsrepräsentanz*) in the psychic domain. Henceforth, this entrance bestows upon it a status ordered in the form of primal phantasies (castration, seduction, primal scene) qualified by some as phylogenetic schema (Sztulman, Barbier and Caïn, 1986). These primal phantasies are all theories advanced by man from earliest childhood to account for the enigma of his origins. Castration binds our existence to finiteness in the form of the fear of losing the detachable object of the body that is our attribute of sexual lust. The phantasy of seduction translates our incestuous desires, mixed with fear and terror, accusing, rightly or wrongly, those close to us of having had the intention of deriving enjoyment from our body. And the primal scene is the unconscious history that we recount to ourselves about the sexual relations of our parents as the origin of our arrival in the world. Tragedy – or its modern form, trauma – occurs when external reality and the factual give rise to the realization of these phantasies, in spite of the fact their status forbids putting them into action. It is to the extent that they are impossible to realize that they are operative in the psychic structure of the human being. These phantasies are an integral part of the human heritage as mythological and universal entities.

To return now to the relation between drive and instinct. Contrary to instinct, the drive does not possess a momentary but rather a constant force of impact. This constancy is regulated by a particular mode of temporality that may be designated as the onset of drive activity (*dé-buter pulsionnel)* that is, an onset (*début*) that is disallowed (*débouté*) over and over again. The *onset* in question is at once repression, the act of rejecting a demand, of *disallowing* a drive exigency, and the return of the repressed, that is to say, what constantly recommences over and over again. The drive is what begins (*débute*) in order to reach (*aboutir à*) a goal (*but*) that it has fixed for itself, except that this goal is on each occasion *disallowed.*[4] For Freud, the drive is *what takes hold of us and what never ceases to take hold of us.*

The discovery of the phenomenon of imprinting (*Prägung*) in ethology turned the Freudian theory of anaclisis (*Anlehnung*) upside down in its turn. According to this theory the choice of the object of love and attachment in the human infant is not a spontaneous and immediate phenomenon. It is by leaning, as it were, on the drives of self-preservation that the infant is able to cathect his objects of choice. As his experience of satisfaction increases, his affective ties towards the mother become associated with the care she provides for him. In other words, for the father of psychoanalysis, satisfaction is first and foremost the pleasure that the baby experiences in the erogenous zones of his body (mouth, lips, skin surface, rectum), which are all seats of his drives. It is this kind of satisfaction that leads him to cathect the external object, the first being the maternal breast. As we have seen in the figure of the partial drive, the object is only constituted through the circuit of this partial drive. Consequently, for Freud, the drive is above all an auto-erogenous impulse based on the vital needs of self-preservation. To the extent that it progressively experiences satisfaction (the back-and-forth movement of the drive circuit), the drive will cathect the object of attachment, namely, the mother (Freud, 1905).

In his theory of anaclisis, Freud is chiefly concerned with the question of psychic energy and its distribution. This is why he places the accent on the autonomous functioning of this energy which, according to him, concerns the vital functions of the baby which underpin the auto-erogenous drives. On this view, the living entity in its solipsism of autosatisfaction comes first, and is followed by the cathexis of the object of attachment.

The discovery of imprinting by Lorenz in 1935 (Lorenz, 1935) formally contradicts the precedence of the living thing over its object. It was not until the 1940s that we learnt that it was possible for a state of marasmus, or even of actual death, to occur in an infant when he is deprived of his object of attachment. René Spitz 's (1945) studies on hospitalism have taught us that in spite of all the care provided for the child, he (or she) may let himself waste away in the absence of maternal desire.

On the strength of the studies carried out by P. P. G. Bateson (1966), W. Sluckin (1965), and R. A. Hinde (1963), following Lorenz's discovery, John Bowlby (1969, 1973, 1980) succeeded in laying down the first stones of the edifice known as *attachment theory*. According to this theory, from the moment of his arrival in the world, the human being is endowed with a schema of dependence (that is, of imprinting) on the principal figure that affords him care and affection.

Attachment theory ran counter to the Freudian concept of anaclisis. It also contradicted behaviourist theory inasmuch as it highlighted the fact that the mother/child relationship depends neither on learning nor on conditioning.

Even if John Bowlby (1969) declared initially that Freudian theory was null and void (p. 173), the Freudian theory of the drives not only survived the phenomenon of imprinting and attachment, but was developed more fully by psychoanalysts in general and by Jacques Lacan in particular – to the point, even, that Konrad Lorenz finally abandoned the term "instinct" in favour of that of "drive". Drive theory no doubt possesses a force of articulation and coherence that ensures its deserved continuity.

What led Lorenz to turn towards the notion of drive was not only its quality of vacuum action, to which I referred earlier, but also the fact that the notion of instinct had very much lost its immutability in the eyes of ethologists. That is why it was no longer opposed to intelligence. Above all, the concept of drive opened up other horizons to which we would not have had access if we had stayed with the notion of instinct alone.

Notes

1 Heidegger distinguishes *Being* from the *existent* (a noun, and not the present particle, designating *what is*). The ontological is to Being what the ontic is to the existent.
2 See further on, Part II, "The thalamocortical model or the temporal loop of oneiric activity".
3 This would explain why James Strachey, the first translator of Freud's work into English, rendered *Trieb* by "instinct" and not by "drive". Subsequently Lacanian psychoanalysts favoured the term "drive" which is more evocative of the linguistic sequence of signifiers than of animal behaviour.
4 See the etymological ambiguity of *but* and *bout* in old French (Bloch and von Wartburg, 2008).

5

THE TEMPORALITY AND PHYSIOLOGY OF MOVEMENT

The specificity of the exegesis, outlined above, of the concept of time in Aristotle, is to highlight the principle according to which time is not reducible to what *is* (*das Seiende*).[1] It differs from it inasmuch as it can be distinguished fundamentally from the Cartesian notion of extension. This obviously does not mean that it is in a state of divorce from the latter. In the Cartesian dichotomy between matter as extension and the soul as thinking, modern science assimilated the latter with the former in order to arrive at a conception of man as a computational machine. Not reducing time to extension may open up other horizons for us in the study of living matter.

In his treatise on time, in chapters 13–28 of the *Confessions*, Saint Augustine (1960) takes up a constitutive element of Aristotle's study which establishes time not only as numbered number, but also as numbering number. If time is something that belongs to movement, according to Aristotle, it is at the same time measurable, a numbered number. According to Heidegger, it is more precisely "the number of movement", that is to say, the number that is numbered in the cadence of movement. However, movement itself requires time for its deployment. That is why Heidegger formulates Aristotelian time as the numbered number of movement. The second hand of my watch gives me the time with its movement. It does so inasmuch as time is the numbered number measuring the movement on the watch face. This is why Aristotle says that time is also the numberer. But isn't an agent necessary for numbering it? Should we consider, then, that time needs a subject? The question is one of knowing if time belongs to the objective sphere of the outer world or to the subjective domain. Even though a thinker like Saint Augustine gave priority to this latter aspect as an answer, the question concerning the status of time still continues to intrigue us. It might be more fruitful to study temporality in the living world in greater depth before returning to the opposition between objective and subjective. That is what I am going to endeavour to do in what follows. As long as the premises of the distinction between objective and subjective remain unclarified, we are in danger of remaining in the aporia that is the dichotomy between the two categories. I think that such a clarification can only be achieved in the course of concrete investigations, while setting aside as far as possible unquestioned assumptions. In so doing, we will discover that there is a current of research in physiology that can open up other horizons for us in respect of the question of time.

This line of research attempts to free itself from the sway of linear time and interests me in more than one respect. Not only does it open up other hitherto unsuspected horizons concerning living matter and its physiology of movement, but it also allows us to envisage a new understanding based on phenomenological temporality as outlined in the preceding chapters.

Movement is what gave rise during the course of evolution to the emergence of the nervous system in living things. Indeed, the danger for animals is to move around in the absence of a programme that is capable of adapting to the vicissitudes of the external world. This is why "the nervous system has evolved to provide a plan, one composed of goal-oriented, mostly short-lived production verified by moment to moment sensory input" (Llinás, 2002, p. 18). Anticipation is indeed a basic act of protection.

Take the example of the reflex which is the simplest and most fragmentary animal movement. It is an automatic act aimed at protecting the sensitive and supposedly vital parts of the body in a stereotyped way. Perhaps it was its involuntary, and primarily automatic character, deemed to be sufficiently protective, that relegated it to the spinal cord rather than to the brain. To protect the body as a whole, however, it is indispensable for there to be a single organ that centralizes anticipation. This organ is the brain whose functioning relies on the economy of energy. This parsimony would obviously not obtain if physiological temporality obeyed a linear system.

Anticipation is a temporal mode that is indispensable for accomplishing movement. The slightest act depends on it. I am getting ready to move a pile of books on the table. I anticipate their weight, the orientation of my body, its gravitational centre by leaning forwards, the strength required, and many other parameters as well, in order to execute the act in question. All these elements of anticipation are simultaneous.

Anticipation not only requires the *retention* of what has just occurred, but also the *protention* of what is likely to follow the act. It possesses a mode of temporality in line with Husserlian phenomenological time discussed earlier. The three temporal moments unfold simultaneously: the present (the intentionality of the subject); the past in the form of retention; and the future as protention of what is likely to happen. In the posture prior to the combat between two felines, each animal remains immobile, while remaining aware of the hostile gestures of the other (retention) so as to envisage the appropriate attitude to adopt (protention), depending on the nature of the next gesture of the adversary. When I am playing with my dog and am waggling the little stick before throwing it, he is watching the slightest movement of my hand, the slightest sign of the direction in which

I am going to throw it, and my posture indicating the immanence of my gesture. Llinas writes:

> To continue with this train of thought, one may propose that the execution of rapid voluntary movement consists of two components with differing forms of operation. The first component is the *feedforward*, ballistic (no modulation *en route*) approximation of the movement's end point (get your hand close to the carton of milk) in which only advance sensory information can be used to shape the initial trajectory of a movement (open loop). In other words, we see the milk carton before we reach for it, and this sensory information is fed forward to the premotor control system to help it choose an appropriate reaching movement we should then make. The second component fine tunes the movement. This component operates "closed loop", meaning that it allows for sensory feedback to refine the movement as it is being executed, using tactile kinesthetic vestibular (balance) or visual cues (get a hold of the carton).
>
> (2002, p. 37)

We can see here how the physiologist employs in his own way terms that are close to those of the philosopher to describe in the same way the temporality of the act.

There are two physiological theories of movement. First, the cognitivist theory prioritizes laboratory research and its approach to living things remains essentially discursive. It dissociates perception from action and sets itself the task of identifying the programmes and schemas that it attributes to brain functioning. According to this theoretical approach, movement is dependent on a centralized command which, like the computer, is equipped with an up-down system.

The second current of thought was based from the outset on studying its object in a natural milieu, *in situ*, without depriving itself, however, of the benefits of laboratory research. According to Etienne-Jules Marey (1830–1904), one of the illustrious pioneers of the so-called ecological theory, "it is not in the ordinary laboratories of physiology that movements can be studied" (Marey, 1895, p. 392). Marey was the originator of the first photographic apparatuses designed to capture movements in images. Moreover, his associate, Georges Demenÿ, sold his cinematographic rights to a certain Léon Gaumont. He chose as his object of predilection the study of movement in physical education. As for the Russian scientist, Nikolai Bernstein, he carried out his investigations in an almost clandestine manner in the ex-USSR. These did not seem in keeping, in the eyes of the Soviet authorities, with the official Pavlovian theory of conditioned reflexes. His fruitful theories and hypotheses continue to stimulate those who adhere to the non-cognitivist movement. The work of Alain Berthoz, at the Collège de France, is particularly worthy of mention here.

According to this holistic theory, perception and action stand in an *intrinsic* and not a dissociative relation to each other. Movement is governed by a nonlinear system in line with the Husserlian conception of time. Every act is what has become known, since Francisco Varela, as an *emergent property*, a concept to which I shall be ascribing particular importance further on in my study of dreams. This amounts to saying that muscular movement does not necessarily depend on an *up-down* system. It only possesses a limited number of pre-established schemas, which modulate its parameters each time in accordance with the constraints of the environment, in order to constitute itself as an emergent property. In other words, it is by borrowing the pre-existing forms that each act adopts new forms.

The brain is dependent on time and movement

Contrary to the traditional conception, there is no sensation without perception. The latter is itself in such close proximity with the subject's experience that it is indistinguishable from the object that it is supposed to target. When the house bell rings, what reaches me is closer to me than the perception that it occasions in me. The reason for this lies in the fading of the subject behind the object of perception. In other words, the subject *is* the perception of the object. Here, as elsewhere, the subject's mode of being is first and foremost anticipative. Without subjective subversion, which is closely bound up with the anticipative mode of perception, we would be in danger of losing sight of the very functioning of the brain. When we attribute the execution of a movement to this or that part of the encephalon, while suggesting that the decision regarding its execution comes from such a network of specialized neurons, it is legitimate to ask the following question: if the person plays no part in the accomplishment of these actions, why then does the brain execute them? If one were to pursue this question, it would simply culminate in an inconsequential and endless regression going from instance to instance, from cerebral network to network . . . To attribute an action to this or that region of the brain is to lose sight of the subversion of the subject. It is because the subject subverts himself behind his action that we mistakenly start looking elsewhere for the provenance of this action.

If all sensation is perception, it has to be conceived of as active, that is to say intentional in the Husserlian sense of the term. Remember that, for phenomenology, consciousness is always consciousness *of* something. Perception is thus an act that is at once projective and anticipative. All investigators in neuroscience today agree that the execution of the action is always preceded by its cerebral activation. Only an inhibition of this activation at the last moment will prevent its muscular execution. This confirms the key idea of Husserlian phenomenology according to which every phenomenon of consciousness is a veritable *act*. The temporal gap between cerebral projection and the muscular execution of the act is of the order of a thousandth of a second. This is a truly internal model governed by the laws of *emulation* and not those of simulation characteristic of computers. "Simulation," Alain Berthoz (2000) writes,

> means the whole of an action being orchestrated in the brain by internal models of physical reality that are not mathematical operators but real neurons whose properties of form, resistance, oscillation, and amplification are part of the physical world, are in tune with the external world.
>
> (p. 22)

Berthoz's formulation is sufficiently nuanced so that we do not suspect these internal models of a disavowed dualism opposing the external world and that of brain functioning. The projective mode in question is constantly modulated in accordance with external variations. The chief task of the nervous system is more one of modulating than of informing. That is why it can function independently of external stimuli without losing sight of their reality and nature. We will see this primordial principle at work in dream activity which, contrary to the common conception of it, is never divorced from external reality. The brain does not need to be constantly in the service of external world. Its function is rather to modulate what it receives from the outside world in accordance with the context in which the latter presents itself. On the other hand, it cannot exist without the uninterrupted movement of the nervous system in relation to a specific mode of temporality that we are going to identify.

For Rodolfo Llinás, movement is the raison d'être of the encephalon. He writes:

> The first issue is whether a nervous system is actually necessary for all organized life beyond that of a single cell. The answer is no. Living organisms that do not move actively, including sessile organisms such as plants, have evolved quite successfully without a nervous system. And so we have landed our first clue: a nervous system is only necessary for multicellular creatures (not cell colonies) that can orchestrate and express active movement—a biological property known as "motricity".
>
> (Llinás, 2002, p. 15)

According to Llinás, one thing is clear: ". . . active movement is dangerous in the absence of an internal plan subject to sensory modulation" (ibid., p. 18). But the movement of animals is not reactive. Thanks to increasingly complex mechanisms, they have become proactive and capable of anticipating external events so as to protect themselves, with the aid of their plans of action, against dangers. "The species that passed the test of natural selection," Berthoz writes,

> are the ones that figured out how to save a few milliseconds in capturing prey and anticipating the actions of predators, those whose brains were able to simulate the elements of the environment and choose the best way home, those able to memorize great quantities of information from past experience and use them in the heat of action.
>
> (2013, p. 3)

For Rodolfo Llinás, the capacity to predict future events is principally the task of the brain. In his view, it marks the highest state of evolution of living beings. He defines prediction as the neurophysiological capacity to foresee what might happen in the immediate future for an animal, taken individually, but also for its species in the longer term. The essence of movement, for Llinás, lies in the fact that it is always oriented in animals towards a precise goal. This is why, he says: "Underlying the workings of perception is prediction, that is, the useful expectation of events yet to come" (Llinás, 2002, p. 3).

The temporal mode of kinetic control

The brain is not in an omnipresent relationship with the external world. Rather, it modulates itself in accordance with its needs and intentions of the moment. It does not let itself be overwhelmed by present tasks at the price of being encumbered by them. Llinás (2002) qualifies this manner of proceeding as the "look ahead function". For him, the procedure in question is the inherent property of the neuronal circuit, for "prediction begins at the single neuron level" (Llinás, 2002, p. 25).

Unlike the computer, the brain does not exert omnipresent mastery over the organs executing movement. Bernstein had pointed out that the control of movement was not a continuous movement but *discrete*; but in order to confirm the validity of his hypothesis, it was necessary to wait for the investigations undertaken by Llinás and his colleague, Ribary, in the 1990s (Llinás and Ribary, 1993). This research established that all voluntary movement rests on constant oscillating and involuntary neuronal activity called *physiological tremor*, which occurs in the muscular fibres at a rate of ten oscillations per second (10 hertz). In the absence of this discrete system of oscillation, one million gigahertz of power would have been necessary, according to Llinás' calculations, to account for all the possible combinations in the execution of the simple action of lifting a small weight of one kilogram.

Movement, and the perception that accompanies it, become possible, therefore, thanks to oscillating and pulsating loops functioning like background noise, whose frequency seems to be invariably stable between 8 and 12 hertz. This rhythmicity, which underlies all movement, continues to function even in the state of rest. Llinás starts out by asking whether this so-called physiological tremor is not designed to endow the nervous system with the capacity to control voluntary movements efficiently, both in terms of moderating their speed and in terms of saving energy. "A relatively straightforward approach," he writes, "to reducing the dimensionality of motor control for the brain is to decrease the temporal resolution of the controlling system, that is, to remove it from the burden of being continuously on-line and processing" (Llinás, 2002, p. 29). To exert this kind of control, it is sufficient to divide up the duration of the operation into small temporal units. This leads us, in effect, to an oscillating and pulsating system that gives rise to a discrete discontinuity. As Llinás points out, this had already been observed by E. A. Schafer in 1886. He had in fact noticed that the curve of muscular contraction invariably indicates a series of successive undulations throughout the whole operation. Sherrington, the great British neurologist, had also noted in 1910 that "the scherzo of Schubert's, Piano *Quartet No. 8*, requires repetitive hand movements at approximately 8 Hertz, which approaches the upper limit for finger movements by professional pianists" (Llinás, 2002, p. 30).

Voluntary movement merely follows the pulsating and discontinuous rhythm that is the cadence of physiological tremor, which is always already in place in order to ensure every eventual muscular execution. The speed of repeated voluntary movement will never exceed that of physiological tremor (8–12 Hz). It has been established that these oscillations are independent of both the velocity of the voluntary movement and the load imposed on the muscles. The rhythm of this discontinuity remains stable even during rest and independently of all active movement. According to Llinás, this oscillatory rhythmicity does not belong to the muscular tissues but comes from the *forebrain* which is pulsatory by nature.

The physiology of movement is not as yet in a position to explain the reasons that govern such a choice of temporality in the management of the dimensionality of muscular movements. The oscillatory and discrete rhythmicity that underlies them can be explained in terms of management economy and parsimony with the aim of preventing voluntary movements from being overwhelmed in terms both of speed and intensity. However, the existence of these discrete oscillations that are operative during voluntary movements as well as in the state of muscular rest is generally linked to their greater efficiency, as a pulsatory phenomenon, in controlling movements. A continuous and linear system would certainly not have been able to manage these with the same degree of efficiency. Llinás seems to give pride of place to the thesis that such a discontinuous system is able to anticipate coming events more easily. But he goes even further in his explanation in order to include the physiological tremor within an evolutionist perspective, whereas other researchers are either perplexed by the phenomenon in question or dubious as to its importance (Berthoz, 2013, p. 22).

The permutation or genesis of physiological temporality

The central pattern generator (CPG) is a neural network in the spinal cord which functions independently of commands from the brain and sensory stimulation. This means that once it has been activated by the brain motor, it can generate motility in the subject independently. In other words, the CPG is the only network that is able to guarantee locomotion without referring to the encephalon. The spinal cord is consequently the sole executive centre of locomotion. We owe this discovery to the British neurologist Thomas Graham Brown (1882–1965), who had the courage to counter the dominant theories of reflexology of the era according to which movement and locomotion were simply responses to external stimuli.

Walking is not an acquired aptitude, but inherent to the animal organism and prior to brain evolution proper. We have seen that, according to Llinás, evolution towards a nervous system requires the prior acquisition of the aptitude for movement and displacement. The genesis of movement is thus independent of the external world. It is nonetheless the case that to obtain the status of behaviour, movement must not only take account of external constraints but also adapt to the requirements of the brain.

The investigations of Graham Brown carried out between 1910 and 1927 did not receive the acclaim that was expected. They were largely forgotten until the 1960s when there was renewed interest in his discoveries. Graham Brown seems to have brought his investigations to a halt as early as 1927 in favour, it is said, of his passion for climbing Mont Blanc. And yet it was in the year of 1927 that he was elected a member of the prestigious *Royal Society*, that is, two years after the sudden death of his father, an influential man of authority and a close friend of Sir Charles Scott Sherrington. By all accounts, Graham Brown's entourage saw him as the second greatest British physiologist after Sherrington (Jones and Stuart, 2011).

Brown began with the observation that animals retain their aptitude for walking, as an organized and well-structured behaviour, even when their afferent neuronal paths have been severed between the lower members and the spinal cord (see Brown, 1912). Thus locomation is essentially generated at the level of the spinal cord, that is, independently of the sensory paths and the central nervous system. Its mechanism consists in the coupling of networks oscillating on both sides of the spinal cord, the activation of one side inhibiting that of the other, and vice-versa. This reciprocal and oscillating neuronal activity functions like a pendulum. So the activation of the network on one side of the spinal cord stimulates the movement of the leg on the same side, while impeding the activation of the opposite side. The right leg then begins, in turn, to function, while impeding the functioning of the other leg. This oscillation back and forth thus leads to the movement that gives rise to walking. In this oscillation that generates walking, we are dealing with a pair of functional elements, the activation of one of which is followed by the inhibition of the other. It is by virtue of their alternating activation that movement is generated. The movement thus produced contains neither a before nor an after, but consists in a coupling of two half-rotations, unlike a linear succession.

We saw earlier that, for Aristotle, time is dependent on movement. According to him, the essence of time resides in the succession between before and after, whereas, in the present case, the movement generated is oscillatory, without a before and an after in the strict sense of the term. And yet it is by virtue of this oscillatory movement that movement proper, that is, locomotion, takes place, giving animals the aptitude for walking. This acquisition of longitudinal movement gives an animal access to a new temporal mode engendering a succession of nows – that is to say, referring henceforth to the pair of the before and the after which, according to Aristotle, are the constituents that "escort" time.

As can be seen from Figure 5.1, it is by turning around its axis that the movement (A) generates another, the movement (B). The latter proves to be a movement of displacement, that is, of locomotion. Thanks to this *permutation*[2] the before and the after are constituted at the heart of time as a linear succession of nows. The precedent oscillatory rhythm is considered to be the degree zero of the movement in which the two oscillating parts merely inhibit each other. This rudimentary action nonetheless contains within it the permutation of the throbbing oscillation in the shifted movement/time of locomotion.

It is by oscillating from one leg to the other that an animal finds itself, like a "mutation", able to move around in space. But this permutation occurs thanks to a change in the modality of time, that of the pulsatory to the linear. The permutation itself is not linear. If we accept Llinás' point that at the origin of the nervous system there is movement, we must also accept that it is this same permutation that engenders the access to linear time.

The permutation that gives access to longitudinal movement thus equips animals with the capacity to flee and to avoid danger, both major acts in the

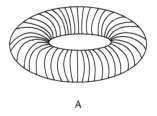
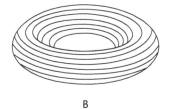

A   B

FIGURE 5.1 Permutation of movement

evolution of living things. An animal is now in a position to adopt a new posture, namely, flight, to ensure its survival. Flight is closely linked to fear which,
for its part, is of an anticipative order. It is not fear, however, that gives rise to
flight. In fact, it is the opposite that occurs in the animal kingdom. Confident
in this capacity, an animal will now be able to develop *in advance* plans and
strategies with a view to executing, in the case of danger, the act of taking flight.
Anticipation thus becomes integrated with an animal's interactions with the
surrounding world. The anticipative fear involved puts the animal into a state
of alert. Freud taught us to distinguish between fear and anxiety, the second of
which has a signal function.

For Llinás, oscillatory movement at the neuronal level is not limited to the rhythmic movement of the spinal cord; it is responsible for the motor apparatus as a
whole. We have seen it at work in the physiological tremor that underlies voluntary
movements. It is also held to be responsible for the synchronization of the constitutive elements of the motor apparatus that facilitates the coordination between the
different premotor signals required for the execution of highly complex movements.
Remember that, for Llinás, the rhythmic movement of the physiological tremor
comes from a cerebral control system that is itself pulsatory. Its discontinuous character intrigues Llinás, who notes that the speed of neuronal conduction responsible
for the control of muscular movement differs from one place to another. He writes:

> Because different neuronal elements (which relate to a movement) cannot be
> physiologically informed ahead of time of the activities simultaneously gen-
> erated by other possible (neuronal) contributors to the final movement, the
> control system must therefore have a clock or a *timing* device. This device
> allows for certain events, correct choices, to be more likely than others.
>
> (Llinás, 2002, p. 44)

Admittedly the spinal cord is capable of any oscillatory rhythm, but it does not
have the means necessary for coordinating all movements at the level of synchronization. Nor is its oscillatory automatism capable of effecting and being
in harmony with complex movements. This is what leads us to be interested in
supra-spinal structures which have the task of coordinating movements and their
modulations.

Several groups of neurons, such as the *bulbar olive,* lead into the cerebellum and play a fundamental role in the coordination of movement. We notice the same oscillatory rhythmicity that we referred to at the level of the spinal cord. In an article published in the journal *Science,* Llinás (1988) also shows the presence in the central nervous system of oscillatory autorhythmic properties which are charged electrically by ionic conduction.

It is at the level of the bulbar olive that the physiological tremor discussed earlier should be located. The bulbar olive serves in effect as a temporal device, like a metronome, for the execution of movements. "The studies of IO [inferior olive] anatomy and function," Llinás writes, "are all consistent with the idea that the IO is in fact operating as a timing mechanism for the rhythmic orchestration of such premotor signals required for the genesis of coordinated movement" (2002, p. 48). The timing of the nerve cells which generates the oscillatory movement does not itself result in movement. A permutation is still required, remember, at one moment or another, to produce a genuine movement in the strict sense of the term; for we know that time is dependent on movement, which produces the pair before/after.

Physiology of initiating action

The studies carried out by the Russian neurophysiologist Nikolai Bernstein (1896–1966) have their place, quite naturally, within the current of thought that goes beyond the neuroscientific body/mind dualism discussed in the third chapter. His line of thinking, as well as his relationship to living things, stands in clear contradiction to the Pavlovian theory of reflexes. This is why they were bitterly contested by the overwhelming majority of scientists under the Soviet regime. And yet the study of movement was entirely in line with the communist policy of the country, namely, of making the combination of work and physical exercise the governing principle of social life. In contrast to Pavlov's reflexology, Bernstein's dynamic conception of movement also proved to be in keeping with the dominant dialectical materialism. But from 1948 onwards, when there was a renewal of anti-semitism under Stalin, his scientific activities and his teaching alike came to a standstill, leaving him the leisure, paradoxically, to continue his inquiries into the subject of movement (Maijer and Bruijn, 2007). At the same time, a number of his colleagues found themselves obliged to make their "*mea culpa*" and to renounce their scientific beliefs. This was the case of the famous neurophysiologist Aleksandr Luria[3] (1902–1977) whose research activities finally contributed, in the early years of the twenty-first century, to the genesis of neuropsychoanalysis (Kaplan-Solms and Solms, 2000). During a scientific meeting in 1951, Luria addressed his colleagues as follows: "In my work, I failed to take as my starting point Pavlov's theory of the motor analyser, basing myself instead on the false physiological conceptions of P. K. Anokhin and N. A. Bernstein . . ." (Maijer and Bruijn, 2007, p. 207). This disavowal did not, however, have any negative effects on the friendship between Luria and Bernstein. Shortly after Stalin's death in 1953, Bernstein was once again authorized to attend scientific meetings.

Bernstein's progress in the study of movement and physical activity is marked by three more or less different periods. The publication of his famous article of 1935, titled "The problem of the interrelationships between coordination and localization" (Bernstein, 1967), constitutes the first stage of his original research. In this article he defines coordination as abundant of degrees of freedom in kinematic movements. We can see that he was already going beyond the body/mind dichotomy.

The second important moment in Nikolai Bernstein's scientific maturity is marked by his research in 1947, *On Dexterity and its Development* (Bernstein, 1966), where he sets out to explain how animals that are well developed on the neuronal plane are capable of adopting adequate behavioural measures when faced with unusual situations in order to resolve their problems of movement. Thus an animal immediately finds the necessary solution, one that sometimes remains unique in its entire existence. An animal evaluates the situation first, and then projects itself forwards by anticipating future events; this gives it the possibility, moreover, of learning from experience. It is here that Bernstein's conception of movement as *initiative*, whose consequences the animal alone bears, has its place.

The third stage in Nikolai Bernstein's research unfolded in the 1960s. His last written contribution from this period was a communication in 1966 destined for a conference organized on *The Cybernetic Aspects of the Integrative Activities of the Brain* by the Eighteenth International Congress of Psychology in Moscow (Pickenhain, 1988). This intervention never saw the light of day, however, for Bernstein succumbed to renal cancer before the conference took place. Its planned title was "The immediate tasks of neurophysiology in the light of the modern theory of biological activity". In this communication, Bernstein argues that the theory of the reflexes had now been completely refuted by modern physiology. For him, this theory "does not take account of the most important factors". "Modern physiological theory," he writes,

> puts the main emphasis on the fact that an organism's reaction to a stimulus (both unconditioned and conditioned reaction), in its form and content, is determined not by the stimulus itself but by its significance for the individual. This means that the most important role is played by factors of internal purposefulness, against the background of which an external stimulus is frequently reduced to a trigger signal. In actions saturated with meaning and internal content (so-called spontaneous actions), such a trigger signal may be completely absent.
>
> (cited by Meijer and Bongaardt, 1998, p. 4)

For Bernstein, the task of physiology is to show how the brain manages to produce non-acquired behaviours which are endowed with intentionality and aim. For him, an animal's actions are all deliberate initiatives. These actions take place primarily in animals of a high neuronal complexity. Such animals set about exploring the diverse degrees of their capacities in order to have *stochastic* behaviours that take account of the future in the light of their past and present. "Bernstein aimed at a brain physiology of how animals take the initiative" (ibid., p. 3).

Two almost opposite currents of research

Bernstein is unquestionably the main instigator of this new current in neuroscience that I am designating here as anticipative physiology. We have seen that in the same period, that is, the first half of the twentieth century, Graham Brown adopted a similar position towards Pavlovian reflexology. It was also around the same time that Jakob von Uexküll put forward the concept of *Umwelt,* that is, the world as the animal reappropriates it. According to Uexküll, animal behaviour is not constituted by reacting passively to the stimuli that act upon it, but by integrating itself with its own world (*Umwelt*), a world that it recreates in its own way in order to be able to interact with it. This seems to be in close keeping with the views of Bernstein on the lability and degree of freedom of animal movements. The concept of *Umwelt* could be integrated quite naturally with phenomenological research, which also had its beginnings at the turn of the century with the genius of Edmund Husserl. Inaugurated by Uexküll, ethology was soon enriched by the work of Nikolaas Tinbergen and Konrad Lorenz, who consolidated its place at the heart of the life sciences.

Maurice Merleau-Ponty, who was initiated in phenomenology, gave a colossal fresh boost to studies in human behaviour by exploring the phenomena connected with it, such as perception and the body, in greater depth. His research inspired many physiologists. Today, it commands authority in the domain of the phenomenology of behaviour.

Thus two scientific movements and methods of thought developed side by side in relation to the study of organized life. On one side, there were the neurophysiological studies on movement and coordination. These stressed anticipation as their fundamental aspect and developed in conjunction with ethological research, which henceforth studied animals from a resolutely new angle in conjunction with the flourishing of phenomenology in diverse scientific domains.

On the other side, a new conception of the body and mind emerged in the wake of the invention of artificial intelligence, in particular during the Second World War, and of the immense progress in neurobiological research during the second half of the twentieth century. It was perhaps above all thanks to the digital revolution that this second current of research swiftly imposed itself throughout the world. It comprises several domains that are closely interrelated. First, cognitive psychology, which claims allegiance to the acquisitions of neuroscience, including the study of neurons and membranes at the level of the organization of the central nervous system. This branch of psychology was founded on a highly abstract approach which tended to exclude any form of subjective dimension and elaborated an increasingly sophisticated methodology in order to rival the "hard" sciences. The current of research in question comprises three areas of studies: the *information processing* models, the *neuronal* sciences, and *cognitive* studies which were still grappling with the stimulus/response model as well as with the feedback model derived from cybernetics. Anticipative physiology derived its momentum precisely from what brought it into opposition with these models. To this end, it began by contesting the idea that the *tabula rasa* could characterize the living world.

Beyond the acquired and the innate

Konrad Lorenz's concept of *hereditary coordinations* provided the cornerstone to the conception of movement as the ensemble of synergies that an animal draws on each time from its own repertories, without having acquired them from the external world. The existence of these repertories confirmed without any question the a priori categories that transcendental philosophy had put forward since Emmanuel Kant. The new science of ethology seemed to follow closely the fundamental maxim of Husserlian phenomenology, *zurück zu den Sachen selbst* (return to the thing itself), thereby distancing itself from the generalized tendency in the cognitive sciences which, by trying to do without philosophical concepts, found they were even more trapped by their own assumptions.

On the strength of its hereditary cerebral organization, an animal makes use, in all its slightest movements of the repertories it is endowed with in order to discover the external world. In other words, it is thanks to that which pre-exists experience that an animal never loses sight of the reality of the external world; moreover, it is this that leads an animal to predict future events so as to be able to anticipate them. We are obviously a long way here from the stimulus/response model, from its *tabula rasa* assumption and its linear conception of time.

The essential difference between the two research trends lies, in fact, in their conception of organized life, since the other points on which they diverge stem from this. The cognitivist current is concerned from the outset with the phenomenon of learning, inherited from behaviourism. It is in this respect that the conversion of the behavioursists to cognitive psychology took place with disconcerting speed during the 1970s. The latter's model of artificial intelligence was in close keeping with their interest in learning which had been the fundamental issue for Pavlov from the beginning of his investigations. These were based, however, on a restricted reading of Darwinian theory. According to this reading, evolution is a step-by-step transformation based on the technique of trial and error. It is therefore linear. There is no doubt that we are faced here with something that is implicit but that does not dare speak its name. This implicit discourse claims that any reference to another type of temporality would risk taking us back several centuries and awakening, like a return of the repressed, the old demon of predicting the future, making us fall again into the trap of eschatology. And yet it was by highlighting the need for animals to anticipate dangers that the followers of the opposite trend defended their research into the capacity of animals to predict future events so as to be able to protect themselves against all risks and dangers.

The flesh, the reverse side of a split

One would argue that the notion of the linearity of time involves an element of denial on the part of the scientist. So that rejecting it would be an anti-scientific act that would risk making research depend on what is unforeseeable, aleatory, and uncertain. But the reality is quite different, because by modelling the study of animals on that of the inanimate object of the hard sciences, one is on the contrary preventing the scientist from envisaging other types of temporality at work in the living world. "I propose," Llinás writes, "that this mindness state . . . has evolved as a goal-oriented device that implements predictive/intentional reactions between a living organism and its environment" (2002, p. 3). An animal's faculty for anticipation runs through Llinás' entire neurobiological work.

The cognitivist movement of research conceives of the body in terms of connections and algorithms, while for the opposite current the body is always intrinsically in movement, even during rest. We saw earlier that oscillation is the very basis of the physiological *body*. This incessant movement makes itself flesh and, in this respect, is already mind. As Merleau-Ponty (1964) says, it is as much seeing as seen, as much feeling as felt, as much touching as touched . . . "But my seeing body," he writes,

> subtends this visible body, and all the visibles with it. There is reciprocal insertion and intertwining of one in the other. Or rather, if, as once again we must, we eschew thinking by planes and perspectives, there are two circles, or two vortexes, or two spheres, concentric when I live naively, and as soon as I question myself, one is slightly decentred with respect to the other.
>
> (p. 13)

Here, Merleau-Ponty gives a superb description of what Lacan has taught us about the Klein bottle (see Figure 5.2).

This bottle is a two-dimensional surface which, for the convenience of tridimensional representation, has been given a supplementary surface. We can see how the "interior" space is simply the extension of the "exterior" space, and vice-versa. In other words, what we have is one and the same surface where there is neither an inside nor an outside. For Lacan, it designates the subject as the Cartesian *res extensa*. Let it be said in passing that it is precisely to avoid falling

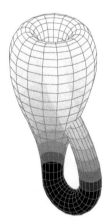

FIGURE 5.2 Klein bottle

into the irreducible dualism of Descartes that Lacan does not identify the subject with the *res cogitans*. In so doing, he follows Merleau-Ponty's conception concerning the indivisible dimension of body/mind as *flesh.*

The Klein bottle offers a marvellous description of the position of the subject of the unconscious as he appears in dreams. In the second part, we shall see, thanks to Lacanian teaching, how the subject comes, in this *other scene* (*andere Schau*), to find himself before the spectacle of the world of which he himself constitutes the backstage. I will venture to show that the subject is never, either when he is dreaming or awake, cut off from the external world. It suffices here to highlight the merit of the Klein bottle in teaching us how a self-crossing surface could lead us into the error of attributing it with two separate spaces, one inside and one outside. This is how the subject/object dichotomy turns out to be null and void. A case in point is the ethological concept of the *Umwelt,* which integrates the specific world of the animal with the very heart of the external world. We are dealing with one and the same world, whose underside is no other than the topside. I will come back to this later.

The body, unity or dispersion

The mind recoils from isolated entities. According to Lorenz, animal behaviour is never an act in the singular. As for Bernstein, he insisted on the *synergic* character of movement. Hence the major object of his research was the concept of coordination.

Husserl built his phenomenology on the principle that every psychic phenomenon is an act. The physiologists of movement found in this maxim the echo of their own object of study. It is in the act that the object of the subject's intentionality *is constituted.* By intentionality, Husserl, following Brentano, meant that each act is intrinsically related to an object. One cannot speak of a subject on one side and an object on the other. It is by means of his object that the subject succeeds in constituting himself. This key idea was radicalized further, giving rise to the subversion of the subject in Lacan's theory. Far from being the cause, the subject turns out, rather, to be the effect of his object. He is nothing but division and fading. The object, Lacan says, is the cause of desire. It divides the subject who asks himself where his desire comes from, from himself or from the Other?

Life is not an abstract concept; nor, on the contrary, is it an entity confined to a laboratory. The biological conception of life is a naturalist interpretation of the Greek βιος which was certainly related to action and production, but also to the individual's *mode of existence.* We touched on this earlier in the form of the "object-little-soul" (*object petite-âme*). The animal is "sick" from life, just as we say that someone is sick from love. But there is more; life is not just an animate entity that is given to my organism. It would be more adequate to speak of *world,* the ineffable world of the real (*réel*) that gives rise to life, the hidden world of the desire of the Other which divides me; and, finally, the world as *Umwelt* in which I move insofar as I am moved but also moving.

Merleau-Ponty spoke of flesh. The flesh is the body that is felt but also feeling, seeing and seen, touching and touched. It is the stuff of the subjectivizing ego, but at the same time the objectivized ego. Merleau-Ponty (1964) conceives of it as a unity that sutures the inside and the outside, like "a being of two leaves" (p. 137). The body, according to Merleau-Ponty is undoubtedly flesh, but also *chiasm* (crossing-over) between subjective and objective experience. As such it harbours antagonistic forces. It is at once identity and difference, silence and speech, pleasure and displeasure, joy and suffering. He writes:

> The axis is alone given – the end of the finger of the glove is nothingness – but a nothingness one can turn over, and where then one sees 'things' . . . The only 'place' where the negative would really be is the fold, the application of the inside and the outside to one another, the turning point.
>
> (Merleau-Ponty, 1964, pp. 263–264)

Flesh is thus at once unity and division.

Dispersion is inherent to the body. That is why the drive is irreducibly partial. Flesh as chiasm combines dispersion and assembly, integrated body and fragmented body. The instinctual drive body is inexorably partial. Would it be a mistake to speak of the body as united? Merleau-Ponty constantly characterizes the flesh by oxymorons. Is the lived body an implacable unity or a fragmented ensemble? Does the subject perceive it in a united or dispersed form? Take the example of the narcissistic body formed between the age of 6 and 18 months. In effect, this constitutes a revenge on the child's fragmented body grappling with his partial drives. Unity and dispersion are thus lived in turn in an inverted form; this is the chiasm of the narcissistic body.

Diversity of the mechanisms of anticipation

Anticipation is not a single centralized system at the level of brain functioning. The sensory captors are themselves endowed with anticipative capacity. The best example is what are called *saccadic eye movements*, changes of direction of the eye when it turns on its axis in order to follow the shifting visual object. The speed of saccades varies between 20 and 150 milliseconds. The shifting of the gaze from one object to another does not affect the stability of the surrounding world. This is due to a set of eye mechanisms, including anticipation. "Evolution made the brain," Alain Berthoz writes,

> a machine for predicting, not just a machine to take account of situations. It also made of it an organ for detecting, predicting, and interpreting movement, for there is no action without movement. Gaze *orientation* was one of the first functions that required the development of a brain that could predict, a brain that was curious, and a brain that could simulate action.
>
> (Berthoz, 2000, p. 181)

Anticipation is equally in operation in the receptors responsible for the diverse regulatory elements of bodily movements such as speed, acceleration, slowing down, shocks, and avoiding collision with the surrounding objects. These regulatory elements make it possible to anticipate the future position of the body as it moves around. "The neuro-muscular spindles measure the speed of muscle stretch . . . when we are going to execute a movement, the response of the spindle can be modulated by anticipating the movement itself" (Berthoz and Petit, 2006, p. 64). The perception of distance as well as the positioning of the limbs are other anticipatory means. The body has a tendency to determine distance by the measure of time, that is to say, the duration of events.

It would be beyond the scope of this book to give an overview of the means of anticipation that the body has at its disposal for executing its movements. I will just mention the reflexes I discussed earlier. These obviously count among the most elementary means of anticipating the dangers that are a potential menace to animals.

Decision-making is another theme of predilection in neurophysiology. As the cognitivist tendency is in the majority, approaching the theme in question by means of qualificative categories, such as feelings and emotions, had remained almost prohibited until recent times. It was the famous neurologist, Damasio (1994), who succeeded in breaking new ground by asserting that decision-taking is not only a rational act but closely linked to neuro-qualificative phenomena. These are essentially affected by the subject's past experiences which serve him as a basis for all aspects of anticipation involved in his decision-making. Such a discovery could have turned the whole rational field of cognitive research upside down. But this was not the case. Quite to the contrary, this research was swiftly made available to company executives who succeeded in interesting the market as well as those in charge of human resources. Their aims were to gain time and increase the *predictability* of events in order to increase their profitability (Kenagy, 2009). Computer software programs were made use of on a massive scale to assist or even replace financial decision-makers (Sprague, 1980).

The question concerning the ins and outs of the concept of decision-making is a vast domain of research comprising different disciplines such as mathematics, cognitive psychology, games, neurophysiology, and, of course, philosophical thought. In his theory of causality, Aristotle distinguished four different categories: material, formal, efficient, and final causes. He added two accidental causes under the terms of *fortune* and *chance* which intervene in one way or another in our decisions. I will discuss these two Aristotelian categories in connection with dream activity in the second part of this book. In his magnum opus, *Being and Time*, Heidegger (1927) gives a prominent place to the question of decision-making as one of the emblematic modes of man as being-in-the-world. *Entschlossenheit* – which might be rendered by *de-liberation* – designates for Heidegger the decisive moment when the *Dasein* – man insofar as he is in an irreducible belonging-together with the world – decides to *liberate* that in himself which is most specific. Henceforth he opens himself in the most singular way to himself and to the world. We will see that the same phenomenon could govern the formation of dreams so as to give rise to decision-making during waking life. Anticipation is truly the fundamental element in all decision-making.

The very recent book by Alain Berthoz, *La décision* (Decision-making) runs counter to the dominant cognitive approach which seems very far from penetrating such a complex question as that of decision-making. Concerning the fundamental rationality of the cognitivist approach, Berthoz (2013) declares:

> It is stiff and rigorously cold. It is indifferent to the soft mist of uncertainty; it protects itself against the traps and marvels of the imagination; it would have us believe that the world can be subjected to calculations, that the Vietnam war can be won by the Pentagone's computers.
>
> (p. 8, translated for this volume)

According to Berthoz, the decision to execute a movement requires three heirarchical levels. First, his experiments confirm for him the validity of the theses of Bernstein and Graham Brown concerning the relative autonomy of the spinal cord in the execution of a certain number of movements. We learn once again that the body follows the shortest path, the least complex and the least costly in terms of economising energy in its functioning, which stands in visible contrast with the logarithmic approach of cognitivist physiology. "When we reach for an object with the hand, the angles of the arm, the forearm, and the trunk are linked by a very precise relationship. It suffices, then, to control *a single variable* to master movement" (Berthoz, 2013, p. 143, my emphasis).

The second level concerns the cerebellum, which, like the spinal cord, possesses its own global internal models. As we have seen already, the control of movement at this stage is discontinuous, discrete, and intermittent. It is the same oscillations, said to be physiological and given prominence by Rodolfo Llinás, that underlie them. The third level is found at the level of the ganglions of the cerebral base and cortex. "Thus, little by little," Berthoz concludes, "an internal model of the acting body emerges" (p. 144, translated for this volume).

Furthermore, laboratory experiments have shown that the way the brain deals with pain differs fundamentally from what is suggested by textbooks of physiology. Faced with pain, the different muscles organize themselves in *synergy*. Thus the reaction to pain is from the outset a complete act. It is not reducible to a simple stimuli/response system. The body interacts in an absolutely flexible way. It can even intermittently seek support from other regions that are not concerned by the localization of the pain. This amounts to an anticipative strategy that is organized according to the circumstances of the stimulation and the momentary position of the organism.

Here, as elsewhere, Berthoz defends his main thesis which requires the brain and the world to be in permanent relations of adjacency. The brain and the external world constitute themselves reciprocally. This confirms the Berthozian thesis according to which the brain is the emulation of action, which amounts to saying that even in the real absence of the world, the brain never ceases to be in conformity with it. We will see that this same principle is at work in dream activity. Once again the affinity of Alain Berthoz's thesis with Husserlian phenomenology is noticeable.

Notes

1 Heidegger distinguishes, remember, being (*Sein*) from what *is* (das *Seiende*), their adjectives being respectively *ontological* and *ontic*.
2 The permutation in question is a noncommutative transformation which does not, however, constitute itself as a quaternion.
3 During an exchange of letters with Freud in 1922, Aleksandr Romanovitch Luria informed the latter about the founding of the new *Psychoanalytic Society* in the city of Kazan, while hoping for official recognition by the father of psychoanalysis. He then began an intense collaboration with the famous *Internationale Zeitschrift für Psychoanalyse* until the year 1927 (see Kozulin, 1984).

PART II

Dreaming and sleep

6

THE ONEIRIC PROCESS

The status and place of the dreamer

In the second part of this book we are going to turn our attention to dreaming on both the clinical and theoretical level. We will see that oneiric activity is a singular expression of temporality in the subject, and that it follows on from the phenomenological question of time as anticipation.

First, though, we must clarify the status of the subject in oneiric activity. It goes without saying that this question is different from that of how the subject appears in dreams, which we will be discussing throughout this part of this book.

Who is it that dreams? Who is the agent or the maker of the dream? Freud describes the oneiric theatre as the other scene (*andere Schau*). We will have to determine the nature of the alterity that is inherent in the dream. As for the brain as an organ of oneiric production, the question is one of knowing what its place and function is in establishing the dream. There is a great risk of confusing dreaming and sleep. The dreaming subject differs from the organ by means of which the nocturnal activity is produced.

Remember, first and foremost, that the subject is determined by time. In this respect, he cannot be reduced to the Cartesian notion of extension, but he is, however, *consubstantial* with it. In his irrevocable dichotomy, Descartes distinguished the subject from extended matter without allowing for the possibility of relations between them. This problematic issue "exempted" modern science from inquiring into the intimate nature of the subject by equating him purely and simply with the Cartesian *res extensa*, the major philosophical idea describing the living being as an animal-machine, an idea that was concretized in the invention of artificial intelligence.

Trapped by such a conception, neurology is unable to form a pertinent idea of the subjective functioning of the brain. From phrenology, where it took the morphology of the skull for zones of cerebral exploration, to its current attempts to take the neuronal regions for mental functions, including its attempts to localize the latter in areas of the brain, neurology is constantly grappling with the Cartesian notion of extension. And yet, here and there, we can find research studies that are trying to replace this notion with the concept of temporality. The recent book by Rodolfo Llinás (2002) places the question of time for the first time at the heart of these investigations. In this book, the author is searching for a neurophysiological structure responsible for the circumstances which would permit the individual to recognize himself as the agent of his acts and gestures. We can see, then, that in the absence of conceptual tools, the scientist is scarcely ever able to avoid the pitfall that consists in confusing the different registers concerning the question of the subject. Llinás does not attempt to give a precise meaning to what he calls the "I" and simply retains, without further ado, the ordinary sense of the word which is close to what might be called the individual ego. And so he is unable to distinguish the ego from the I, the individual from his mind, the infatuated self from the subjected subject, sameness from the identical, or the person from his attributes. At the end of his research, he concludes that, in the encephalus, there is no place for, or sign of, the "I" as "first person singular", as actor of its neuronal vicissitudes.[1] And yet it is precisely here that the question of the subject could be deployed to its full extent. I will therefore attempt to examine more closely whether the neurophysiologist has not, unwittingly, been concerned precisely with the subverted subject.

Is the subject a thinking substance, the primordial cause of our acts, an emergent interiority of the brain, an incarnated presence, an essence, a form, a consciousness, an ego, a me or an I . . .? The concept of the subversion of the subject is an attempt to get beyond the aporia of the question.

Referring to the first person singular, the concept of the subject dates back to the Cartesian cogito which gives it an ontological status: "I think, therefore I am". It is an inaugural act which represents wonderfully the beginning of modern times when man became the measure of everything. Admittedly, the archeology of the subject (see de Libera, 2007) dates back to ancient philosophy, when Aristotle put forward the specific term of *hypokeimenon* to designate the substratum, that is, the foundations, of every entity. The concept was transformed in the Middle Ages into *subjectum* (literally, "that which is thrown under"). But, owing to its reference, since the Renaissance, to man, the subject has taken on a radically new meaning. The scholastic "subjectness" (everything has a foundation) has since turned into subjectivity (man is the measure of everything). Let us try now to examine this new signification of the term.

When grappling with his subjectivity, man must now answer the question: Who am I? We can see then, that the ontological (the "I *am*") engenders the existential (the "*Who* am I?") leaving the human being to his aloneness in the world which henceforth remains outside him. Subjectivity therefore comprises not only an existential, but also a moral dimension exhorting man to answer the question: "Who am I?" It is through such unavoidable questioning that the subject discovers himself as the Other. Admittedly, the question "Who am I?" designates the condition of possibility of man; but each time he tries to answer it, he is struck by surprise, if not mutism. We are quite unable to say anything about who we are. Consequently, anxiety is the token of the subjective division in which the "I" cannot but find itself as the Other. This suggests that it is possible to distinguish subject and subjectivity. The first is our condition of existing and, as such, remains the instance of our division; whereas the second is the way in which we relate each time to our condition.[2] The first remains in its division, leaving the second to bear the uncertainties of historical, cultural, social and individual events.

What determines this condition of existing that is the division of the subject? This is where Descartes' metaphysical error comes into play. By ascribing it the status of a substance, the philosopher masks its true questioning value. To conceive of the instance of division as a substance would be an incongruous response to the question. For the division of the subject is related to the cut of the Moebius strip (see Figure 6.1). It only exists to the extent that it gives rise to the two entities for which it is only passage and transition.

Insofar as he is divided, the subject cannot exist properly speaking. How could being be refused to an entity that constitutes man? This brings us back to the question of time. As we saw in the first part of this book, time cannot be reduced to being as *res extensa*. We have also accepted, with Aristotle, that the now is the paradigm of time. Without it, there would be neither a before nor an after, neither a yesterday nor a tomorrow, neither an anterior nor a posterior. It constitutes the *division* inherent to time.

The now is an incessant division between the past and the future and, like the Moebius strip, has a circular form. Beginning with the present instant, it heads towards the past, then in the direction of the future, before coming back in a loop to the present. Time is this infinite prolongation constituted of nows in the form of loops. It follows a circular form in which the loops succeed each other without ever

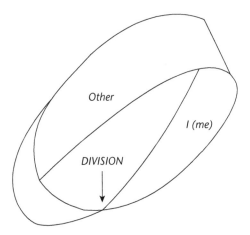

FIGURE 6.1 Moebius strip (division of the subject)

resembling each other. These temporal loops are of a retro-protentional nature, linking the past to the future and vice-versa. Thanks to the now that is the irreducible division of time, the subject projects himself into the future from his past. In this simultaneous operation which Husserl calls retention/protention, the subject embodies the now linking up in as many loops the past and the future. Though divided, the subject can only experience himself factually as a continuum of these nows which give him the feeling that he is always the same. This is what we call the sense of being oneself.

Owing to its division, the now is an instant of time that is always *stretched* between the past and the future. It links *retention,* Husserl says, to *protention.* It embodies this *tension* that is the subject between the self and the Other. If the subject wanted to choose himself, he would lose the Other, who serves him as a warrant. By favouring the Other, he would lose himself. He can only project himself from the desire (*expectation*) of the Other towards whom he is always stretched out.

The now (*maintenant*) is what safeguards (maintains) time. It is the *extension* of itself insofar as it is in its retro-projection that the before and the after of time are constituted. These two temporal forms are in a state of invitation towards the now, which projects itself as expectation towards them.

Without expectation, the subject could not embody this hand *stretched* out towards the past and the future. Is it a passive or an active expectation? Does the subject suffer the passing of time? Or does he constitute himself thanks to the temporal tension between yesterday and tomorrow? Is he the one who causes the relation between them and has power over their *extension* towards each other?

On closer inspection, the expectation that is constitutive of the subject is neither passive nor active. True, the subject is factually and constantly in one state or the other. But there is a more original expectation that accompanies him from the moment he comes into the world until the last moment of his existence. It is not an expectation that concerns a project or a precise intention to be carried out or that depends on a mode of relating to the past. It is more a *state of expectancy.* This makes the subject an expectant being without there being a precise object. His expectancy unfolds in the mode of *awaiting . . .* , an expectancy without object or measurable duration. It is not an expectancy that takes place in the flow of time, but which constitutes its very condition. It is by virtue of it that time falls into its tripartite division of past, present and future. It underlies the temporal flow. It does not exist ontically. In truth, it itself is *time*; the rest is purely *temporal.* It is strange to speak of a time that is not temporal. What is at issue is definitely not a this or a that; nor is it an objectifiable, and even less, a calculable time. We will soon see whether this fundamental mode of expectancy is attested by neurophysiology.

In what follows, we will be examining the temporality at work in dreaming, which is a major projective act that dramatizes or stages unconscious impulses. As we know, the latter are dependent on the drives which we discussed in the first part of this book. But I shall be sketching out a theory of oneiric activity which unfolds as if it were the reverse of the drive; this will be an opportunity to return to the question of the unity of the subject. We will see that the original expectation that I have just been sketching out underpins yet another mode of temporality of which dreaming is a major expression.

At each stage of my research, I will draw on psychoanalytic clinical material to illustrate my remarks, but also on the investigations of neurophysiologists in the domain of sleep.

The temporal mode of desire

For Freud, dreams are wish fulfilments. The specific mode of this fulfilment is temporal. Oneiric activity aims, in effect, to put in *the present* what is *to come*, and it does this on the basis of the past as *having been.* By *having been*, I mean that which still persists, as experience, in the present. *"Das Präsens ist die Zeitform, in welcher der Wunsch als erfüllt dargestellt wird"* (Freud 1982, p. 511): "The present tense," Freud says, "is the one in which wishes are represented as fulfilled" (1900, p. 535).

The insignificant details of the day preceding the dream, called day residues, seem to be at the origin of dream formation. However, they are only apparently trivial *wishes*, and are different from desire insofar as it constitutes dreams. Here desire and wish differ, the latter always being likely to engender desire. A wish is, so to speak, factual; it is just a memory referring to a psychic or external event. As for desire, it pertains to an expectation of quite a different order. The wish of a day residue is *event-based (factual)*; desire is *anticipatory.* A wish is likely to be satisfied; as for desire, it is not a matter of satisfaction but of fulfilment.

Let us see how Freud accounts for this expectation inherent to desire (*désir*). Taking up his dream of "Irma's injection", he speaks of the optative mode of desire as it is *fulfilled* in the dream. "There the dream thought that was represented was in the optative"/"*Hier ist der zur Darstellung gelangende Traumgedanke ein Optativ*": "If only Otto were responsible for Irma's illness!"/"*Wenn doch der Otto an der Krankheit Irma Schuld sein möchte!*" Let us examine more closely dream functioning in the optative. "The dream," Freud continues, "repressed the optative and replaced it by a straightforward present"/"*der Traum verdrängt den Optativ und ersetzt ihn durch ein simples Präsens*": "Yes, Otto is responsible for Irma's illness"/"*Ja, Otto ist schuld an der Krankheit Irmas*" (Freud, 1900, p. 534).

In the same passage, Freud notes another characteristic of the dream, namely, its *visual representation* or *depiction (Darstellung* in German, *figuration* in French), which is responsible for the distortion inherent in dreams. Given that visual representation is the almost exclusive means of dream expression, the words and linguistic devices of dreams are expressed by being transformed into *images*. In other words, hearing is transformed into the visual. This is what Freud calls "representability" (*Darstellbarkeit, figurabilité*).

We may consider that dream depiction (*figuration onirique*) and the suppression of the optative involve the same mode of functioning. Images and depiction are the indispensable means of the dream work. How can a wish, that is to say an optative clause ("if only . . ."), gain access to the image and transform itself into a pictorial form (*figuration*)? The solution would obviously be to suppress its optative aspect. The depiction and the image can only, consequently, assume the indicative mode of discourse ("Yes, Otto is responsible . . .") to express the wish in question.

The etymology of "time"

What about the meaning that language confers on time? In Indo-European languages, time is expressed by two primary notions: "stretching" and "cutting". The Lithuanian *templn* (tighten by pulling) and *tempt-yva* (string of the bow) express this clearly. The German *Zeit*, the Dutch *tijd*, and the English *time* derive their origin from the same semantic source as the Latin *tendere* (stretch, extend). In Greek, τενσω is akin to the Latin *tondeo* (shear) and has the sense of cutting and cutting up contained in the etymon *tan-*.

On another level, the Latin *hora* (hour) – in Greek ώρα, and in Old Iranian *hour* (sun) – is very probably at the origin of the Greek άρκτος (arctic).[3] Whence ουρανός, which means starry sky. The same etymon is at the origin of the Latin *arcus* (bow, arch).

Thus, the etymology of these terms designating time refers to the idea of stretching and cutting. Time is henceforth what lasts, like a taut string, like a bow that endures the movement of tension until the ultimate moment of the "cut" when the arrow suddenly flies away. It is this tension and tugging which give rise to the propicious and extreme moment of the event. The Greek χαιρός has the same meaning, that is, the propitious moment, as the German *ziemen* (to be proper, suitable) and *ziemlich* (fitting) derived from *Zeit* (time). *Heure* (hour) in French, as well as its derivatives (*or, désormais, lors, dorénavant*), designates essentially the [un]timely" ("*[in]tempestive*") character of time. Language gives prominence to a mode of temporality tending towards a point of tension which finally produces the event. What determines time, therefore, is not a present that passes from moment to moment, equal to itself in its self-succession, but what is governed at once by expectation and the non-actual. In the now (*maintenant*) of the present, this characteristic of tension is constantly main-tained (*main-tenu*).

The temporal mechanism of the dream

Let us return now to the time of desire that depends on the temporality of the unconscious. The repression of the *optative* in dream activity is an attempt to incite desire to fulfil itself in the *indicative* mode. The repressed desire, that is, the optative, is matched by the manifest content of the dream experience that has assumed the indicative mode. Remember that the dream has two types of content: one *manifest* (the narrative as given by the dreamer) and the other *latent* (its meaning that is hidden in the meanders of its unfolding). That is why dreams require interpretation. The gap between these two types of content partakes of the dreamer's sense of division.

The optative in dreams is not simply an "I wish that . . .". Its temporality, governed by expectation, distinguishes it from every other subjunctive discourse.

The expectation in question is to be distinguished from any form of fortuitous or event-based expectation. The human subject does not wait for his turn in a queue in the same way as he deploys his desire in a dream.

Time for man does not receive its determination from the past, but from the future. The latter is not a simple temporal entity in front of the subject so that he orients himself towards it. However paradoxical it may seem, the future for man precedes every other temporal dimension.

Freud distinguishes realization and fulfilment (*Wunscherfüllung*) of desire. Realization exhausts itself in events. Fulfilment, on the contrary, involves taking into account the non-actual; it is in a position to welcome – we will see how – surprise and the unforeseen.

Far from being the realization of a project, the future in question is the projection of the present time as if fulfilled. From now on I shall call this future by the term "*advent*" (*avent*) and the temporal process that governs it by "*adventization*" (*aventisation*). It goes without saying that I am borrowing this word from the Christian tradition which confers on it the sense of what is *imminent* as future. The advent is not measurable time. It does not flow as such in time. The present is not a now that advances towards the future as advent; rather, it is the latter that goes to meet the present and determines its mode of temporal flow. The advent is a particular kind of expectation that is accomplished in the division of the subject who is divided between a present that tends towards the future and an event that is awaited. Adventization embodies this tension which disposes the subject to give himself the time needed to traverse his desire laboriously. I will come back to this later.

Freud's inaugural dream

How does the dream, the guardian of sleep (Freud), differ from the *waking state*? Here, *paradoxical sleep*[4] acquires its full significance. Insofar as it is the royal road to the unconscious, the dream is the attempt to *adventize* desire in the sense I have just ascribed to this term. The liturgical tradition understands advent precisely in the sense of wakefulness and vigilance. The fulfilment inherent in the time of desire is the first task of dream activity. It nonetheless remains at the level of an *attempt*. If adventization consists in letting the time of desire come into being, it would not be surprising if the subject succeeded from time to time in foreseeing the events that are constantly signalling themselves to him. The dream of Irma's injection is an illustration of this.

Is it a matter of chance if this dream gave birth to the discovery of oneiric activity as the fulfilment of a wish. "I reproached Irma," Freud writes at the beginning of his account of the dream, "for not having accepted my solution" (1900, p. 108). What, then, is this solution that presents itself with insistence at the beginning of the dream? The essential question for Freud, at the origin of his discovery, is the quest of the hysteric, incarnated in a body moved by the quiverings of her desire. Is hysteria an organic affection or a desiring vacillation? What about the medical corps represented by the assembly of doctors that inhabits the dream stage?

It corresponds to a certain fear: "I was alarmed at the idea that I may have missed an organic illness" (Freud, 1900, p. 109). Giving the body the status of the high place of desire is indeed the challenge that the Viennese doctor throws down to his peers who are present in the dream. Is the discovery of the unconscious as the instance where the desiring dimension of man lies akin to a scandal? "This is a perpetual source of anxiety," Freud continues, "to a specialist whose practice is almost limited to neurotic patients, and who is in the habit of attributing to hysteria a great number of symptoms which other physicians treat as organic" (1900, p. 109). The dream attempts to adventize this challenge. Was not the reproach made to his friend Otto one that Freud would address to himself before it was time to announce his discovery? Does not Freud's desire waver between challenge and reproach? Was Freud not running the risk of causing a scandal by bringing his own desire contained in his dreams into the public domain, albeit in the name of a scientific discovery, in the hope that one day a "marble tablet" (see Freud, 1985, p. 417) might be made of it? The dream is an attempt to answer these questions, an attempt at adventization which permits Freud to remove the veil over unconscious desire.

The narrative web of the dream

Without wanting to enter fully into the ins and outs of the oneiric world, I am now going to give some examples of dreams so that we can begin to impregnate ourselves with their strange and singular structure. The brief interpretations that accompany them are not intended to exhaust their meaning; they are given merely as suggestions as to how they might be understood.

The following dream occured at the beginning of an analysis of a woman of about forty: "I was with my parents; the weather was grey. A snake was coiling itself around my father. Everything was immobilized. I turned towards my mother. Another snake was wrapping itself around her. The two beasts began to follow me. I ran off. Then my cousin arrived. She was not concerned by the danger."

We can notice the speed and the rhythm underlying the telling of the dream. In spite of its briefness, we discover a well-ordered structure which keeps both the dreamer and the person to whom she is telling the dream in suspense. The end of the dream brings some relief to what an instant before was at the height of action. The dream figures seem closely linked to each other by a danger. The narrative is organized around a major signifier, the snake.

The dreamer's associations establish the following facts. When she was born, her mother had to expel the remains of the intestinal parasites from which she had suffered throughout the pregnancy. This "filth" has a deeper reason that has marked the analysand's subjective history. The discovery, during the analysis, of her incestuous desire for a brother who was literally adored by his parents, offended her sensibility. In fact, it is this that is the object of adventization in the dream. The presence of the cousin, her brother's girlfriend, attempts to substitute the unacceptable incestuous desire with the cousin's legitimacy. The skin symptoms that appeared after a trip she had made with her brother could thus be explained before they disappeared.

Now for another example. Here is the dream of an 11-year-old girl which is akin to a fable. "My maternal grandparents were at the top of the Eiffel tower. A giant dinosaur swallowed the whole tower in one mouthful."

The girl's problem was in fact her small size, and her friends would make fun of her on account of this. During a preliminary interview with her mother, the latter, trying to minimize the problem, had claimed that the cause was genetic and came from her own parents. Vengeance, in the dream, towards her grandparents (displaced on to the Eiffel tower), is at the same time an allusion to what had caused the mother to break off relations with them following a family misunderstanding. The real meaning of the dream resides in the working-through by this girl of her wish to remain her mother's little girl. This was why, when her younger 5-year-old brother was born, she began to show signs of rejecting him almost totally. If the giant dinosaur represents her desire in a projective form (her small size), this is, however, merely the sign of the adventization of her *devouring* fusion with her mother, the object of the consultation.

A man was asking himself serious questions about his father's dubious activities during the Occupation. He had the following dream. "In the station of a provincial town, I was sitting on a bench. Suddenly, a man appeared. He met a woman and gave her a message. Three men appeared and killed the man. A person sitting next to me was reading the Libération newspaper. I went to a kiosk to buy a copy. I was told that it had sold out. I went into the town. It was deserted, the shops were all closed."

The dream brings together several memories dating from the 1939–45 wartime period that were either experienced, or of things heard about. It portrays the murder of a father, perceived as a rival, who, in addition, came back after the liberation. But the essential aspect of the dream resides in the adventization of a troubling question, raised by the analysis, which had hitherto remained inaudible for the analysand. It concerned the fact that his father had most probably been a double agent during the war.

In this criminal series, let me cite a last dream. It concerns a remarkably beautiful woman. "A house, my father, my children and my mother. Someone arrived as a sexual object, a girl or a boy, I don't know. We laid him down on the ground. I said that it was a trick. We came to his genitals: 'You see, there's nothing between his/her legs.' 'Look,' my mother said, 'there's a knife sticking in the ground.'"

The dream barely disguises the complex of the little girl, who is without a masculine attribute. It attempts to adventize the lack in question for this little girl who was the imaginary phallus for the entire family. Now that I have given a few examples illustrating the narrative web of the dream, I am going to focus on the different forms of dream narratives.

The diversity of dream narratives

The psychoanalytic apprehension of the dream as text allows us to consider dream activity as having a narrative structure. But what kind of narration is it?

Is the dream structured like a short story, a fable, a legend, a fairy tale or what? The examples cited above show this. The dream can take on any possible form of narration. Certainly the epoch, the culture and the society play a major role in determining this form; but as we shall see later, the psychic structure of the individual also plays an important role in this choice.

It goes without saying that the diverse and varied forms of the dream narrative cannot equal the novel or genuine narration. Neither the duration, the size, nor the issues at stake in dream activity allow us to place the dream on the same level as a literary production. We would be justified, however, in comparing it with what is called a *narrative sequence*. It should be noted that the dream, as an unconscious formation, has the character of incidental speech that erupts in the subjective life of the individual. When one adds to this incidental element the component of intrigue, one obtains what may be called a *narrative sequence*.

Enchaînement, embedding or interweaving

Linguists employ three categories to refer to the relation between two or more narrative sequences. We know that a dream can possess such sequences. Above all they can be distributed in different dreams of the same night. Their relations are determined, then, by three linguistic categories, enchainment (*enchaînement*), embedding (*enchâssement*) or interweaving (*entrelacement*) (See Ducrot and Todorov, 1972, p. 379).

"I was most probably with my wife. We had to go to a room on the second floor of an old house which resembled the building of my primary school with vertically sliding glass windows. The access was by an escape chute leading from the courtyard of the house to the room.

"In the room there was a big table, my wife, perhaps, but in particular a long-standing friend. The table reminded me of winter holidays a very long time ago. The table of the local council of the snow-bound village was indeed impressive. During these holidays, I was prey to feelings of abandonment. I let my travelling companions look after me, which I enjoyed. This seems to be coherent with the presence of a long-standing friend, who, feeling abandoned by his family, became for a while the "adopted child" of my wife and I.

"Now I was leaving the room through a window, which looked like a small skylight. It was difficult to get through. I felt anguished by the idea that I might never be able to get out of it. This kind of situation occurs frequently in my dreams. I must admit that it is akin to a birth. I finally managed to get out of the window and back down the escape chute into the courtyard. I was with my wife again. In the dream we had an adopted boy aged about twenty. I suddenly found myself alone in his company. I was thrilled with joy at the idea that he could come and have lunch with us; my wife was supposed to be at home now. This whole scene unfolded in the places of my childhood. My 'adopted son' couldn't come to our house. He had a date with his girlfriend to whom he was going to offer a gift. I went off on my bike disappointed, abandoned by 'my own' child."

We can see in this dream how the different sequences *are linked up* (*enchâiné*) around the question of adoption, abandonment, and birth. Likewise, the sequence of the "adopted son" embeds (*enchâsse*) each time the sequence concerning the presence of the wife. The *interweaving* (*entrelacement*) of the different episodes brings the "long-standing friend, the primary school and the places of childhood" into close relationship.

The recurrent embedding of the wife by the adopted son highlights the attempt to adventize a specific problem that was worked-through in the analysis. It concerns the maternal position that the analysand has always tried to occupy to the detriment of his paternal position. This issue is overdetermined in the dream by the question of adoption, but also by the wife's recurrent remoteness.

The fantastic and the supernatural

In the dreams that follow we are in the presence of another kind of narrative that may be qualified as fantastic or supernatural.

In his book, *Off to the Side: A Memoir*, Jim Harrison (2002) relates a singular dream. "One winter I was having a particularly severe slump due to exhaustion over a number of items in my life, and at my lowest point I had a dream of God standing in space before time 'was'. He was hurling countless trillions of nearly invisible specks of material into the void" (p. 49). What desire could motivate such a dream? What place should we ascribe at the subjective level to the appearance of the All-Powerful? What can someone who is prey to such melancholy wish for except that which touches the abyssal depths of his torment, namely, death? It is Harrison himself who confirms this for us. "When I awoke instantly – this was the kind of dream that wakes you – I was fearful because I remembered from childhood the Biblical admonition that if you see God you die" (ibid). If it is indeed the sad wish for death that motivates the formation of the dream, one would have to ask oneself by what attempt at adventization it is sustained. To answer this question we could restrict ourselves to the dream imagery, to the Creator, God the Father, engendering the particles thanks to which the human infant is conceived. The challenging ambiguity of the dream consists in adventizing this ultimate term that is death in order to ward it off.

I am now going to mention two other dream sequences touching on the same question of the sacred. They date back to the childhood of a man who has reached an age of great maturity. "I was between 4 and 5 years old when I had these two dreams within the space of a few days."

1. God was hitting my mother with a big stick. The evening before, I had in fact heard my mother blaspheming against the Almighty. So, in the dream, He was punishing her. The strange thing was that God appeared in the form of an elderly, sovereign, and wise lady.
2. Behind our house, in the river dried up by the heat, I picked up a piece of sky in the form of a triangle. It had the consistency of a big piece of jelly that had been polished and flattened out in huge tray.

We do not know if these dreams are screen memories, that is to say childhood memories that never existed but which result from an unconscious phantasy. The analysis nonetheless provided us with a few elements that threw light on the first dream. According to a family legend recounted by the dreamer's mother, the dreamer had fallen seriously ill around the age of 3. Believing he was going to die, the doctors soon decided to stop treating him. Only an elderly woman who was their neighbour was able to "save" him thanks to some medicinal plants. She was his saviour, or at least the one who gave him (back) his life.

As for the divine sanction inflicted on his mother, it may have had a significance other than punitive. There was reason to suppose that it had been an attempt to save the mother from God's devastating wrath. The matter had ended with a simple act of punishment. It is here that the adventization of the dream might be situated.

Oneiric embroidery

In what follows, I will be reporting a certain number of dreams, all of which occurred between two sessions of analysis. They are akin to a kind of narrative that may be qualified as oneiric embroidery. The dreamer is a man over thirty who has suffered from asthma since his adolescence. The symptom had contributed to strengthening his relationship with his mother, a doctor, who, since the beginning of these symptoms, had taken over the treatment of her son. The first step towards real recovery was obviously to take the responsibility for his medical care out of the mother's hands. The dreams below regularly allude to this in the form of rejection, but also of despair, owing to the fact that the dreamer often misses the train whose destination is no other than his mother's home.

"I had gone to see a doctor (a man); I was lacking oxygen. He told me there was no solution. I was going to die slowly." The manifest content of the dream stands in clear contrast with his good mood, mentioning for the first time that he had stopped smoking. The contrast was confirmed by another dream. "I was in a suburban station, feeling irritated and depressed; I was looking for my way." He brought the following associations: "This weekend, I went to visit a friend I have known since secondary school. I often had asthma attacks there. As it happens he doesn't live far from my mother." Suddenly, another scene of the dream of lacking oxygen came back to him, which is a sort of dream arrangement. "But my children didn't want me to die. I went to see my mother to get some treatment. Then I woke up. I had difficulty breathing."

For no apparent reason, he then had the following associations. "For my father, I am the hard one, while my brother is sweet (*doux*), the kind one. I'm always being taken in by his sweetness (*douceur*). In fact, my brother does things on the sly (*fait les choses en douce*)." These associations came after he had mentioned his "incestuous" relationship with his mother. Perhaps they were a projection, for the following dream concerned the mother again, who had henceforth become inaccessible thanks to his understanding of the heavy load of the content relating to their "incestuous" relationship. "I had to catch my train at Montparnasse (maternal destination) at ten to nine. I arrived at "*La porte Montparnasse*" (*sic*), that is, one station after. I was lost. I ran to the station. I missed my train. I was completely desperate." Here is another dream: "I was with my children in a town. I was completely lost. My ex-wife, who avoids me, then appeared. We were still more or less together. I told her that she didn't love me any more. She replied that unlike my brother, I had no desire for her." Here, his ex-wife stood for a maternal imago to permit him to accept/refuse the "incestuous" relationship. Adventization is often marked by such ambiguities; as we will see later, this is due to the relationship of the subject to the real (*le réel*).

The cadence of the dream

The narrative structure of the dream allows us to consider it as a particular kind of intrigue. By intrigue I mean the enactment of the relations between characters and events. Dream symbolism does not lie only in its representability (*Darstellbarkeit, figurabilité*) through which the image becomes the medium of words, but also in the association between its structural elements.

The question that arises is that of knowing what the purpose is of the subject's recourse to story-telling and narration. What is the origin of this need for intrigue in the theatre that is the dream, where desire works its way into the meanders of a narrative sequence?

As a symbolic element, intrigue serves as a third party for the subject. It distributes the places while integrating them within a narrative frame. In his treatment of "The Purloined Letter" written by Edgar Allan Poe in 1844, Lacan (1966b) pins the intrigue of the story around the *objet "a"*, which is *the letter* as cause of desire. It is equally the object, cause of the intrigue of the dream, that conveys desire and articulates the place that falls to each of the characters. *Intrigue is thus the rhythm and cadence of desire.* In this respect it structures the dream. In what follows, I am going to identify some grammatical devices at play in the dream narrative.

The perfective and the imperfective aspect

Literary criticism extrapolates the principal aspects of the grammatical verb, namely, the perfective and the imperfective aspect, from the narrative structure as a whole. They can also be applied to the unfolding of the dream. First, let me explain these two modes of narration: "The French language contrasts the perfective aspect as in, "*l'année dernière j'ai été malade*", and the imperfective aspect as in, "*l'année dernière j'étais malade*", where the sick person is presented as co-extensive with the entire year (hence the stylistic effect of "*l'année dernière j'achetais une voiture*", making it understood that this purchase was the preoccupation of the entire year" (Ducrot and Todorov, 1972, p. 391, translated for this volume).

Dreams seem to privilege the imperfective aspect of the verb. Think, for example, about the beginning of Irma's dream: "A large hall, numerous guests, whom we *were receiving* . . ." (Freud, 1900a, p. 107, my emphasis), which plunges us right away into the oneiric atmosphere and tells to us that the story does not end immediately. The "perfective" dream has the purpose, on the contrary, of distancing the intrigue of the dream and of attenuating its narrative fervour.

The absence of intrigue

Grammatical tenses, then, have a determining character for the intrigue of the dream. It should also be noted that the verb tense of the dream narrative is not alien to that of the dream itself. Depending on the scale of the intrigue and its unconscious content giving rise to the subject's resistance, the grammatical time of the narrative may remain faithful to that of the dream or be added to it subsequently. The apparent lack of intrigue does not mean, however, that it is absent in the dream activity. It forms an integral part of what is being woven at the level of the desire conveyed by the dream. Here are a few examples.

"I had just left a session, not with you but with my last analyst. I noticed that I had left my wallet behind. The building resembled a private residence; there was a large room, like a library. I saw someone coming out of it. I think it was the next patient. My analyst lived on the first floor. I went up the stairs and noticed that it resembled my primary school which was classified as a historical monument. Once I had reached the first floor, I realised I was in the attic. Two men, who introduced themselves as Pierre and Jacques, asked me what I was doing there."

The essential point is expressed at the end of the dream. Knowing that the name of the analyst in question was *Paul*, I understood immediately the dream depiction of the contemptuous French expression, *parler avec Pierre-Paul-Jacques*, which means to talk to whoever happens to come along. Hence the bland aspect of the dream which is devoid of intrigue. "My childhood memories," the dreamer seems to be saying, "are now classified like historical monuments. I find no interest in them. They should be put away in the attic. My presence can no longer be justified. I would gladly leave my place to the next patient." But these words are contradicted by the wallet left behind in the analyst's consulting room. The display of dullness stands in contrast with the patient's constant fear of triggering a catastrophe. This contrast is all the more striking in the dream that followed this one.

"There was a queue of people who were going to be buried alive in their coffins. I was a witness and had to record their deaths." The dream narrative gives rise to a sense of indignation and moral disgust in the subject who accuses himself of all the crimes in the book. "What astonishes me is this heart of ice that I had in the dream. It was a disgusting job, but I did it anyway. What on earth had those people done? They didn't even rebel."

His associations revealed unexpected things. The head-counting evoked for him the *jury of the Pass*,[5] of which he had been a member. "It was a dream without words, just like the work sessions of the Pass where we were reduced to silence in front of the candidates. The people got into their coffins by themselves, the others nailed down the lids of the coffins, and I recorded them. In the sessions of agreement, I did, in fact, sometimes act as secretary."

The unbearable hatred in the dream tries to hide behind the lack of intrigue and the monotony. It is expressed differently in the dream below, where we can see once again that the absence of intrigue is designed to mask the murderous intention of the content.

"I had a sort of watch whose case opened, the sort of ring in which one used to put poison. I would use this round and hollow object to clean my contact lenses. I was constantly putting powder on the skeleton of a dead bird." When I asked him about the last incongruous part of the dream, he distorted the dream about the coffins, reported above, revealing his silent and scornful aggressivity. "You know," he said, "it was like the mouldiness of that dead man on whom I was sitting in the other dream." The dead bird is in fact an allusion to the name of his ex-analyst on whom all the aggressive hatred was focused. Then, he added this association to the title of a book that he had just read, *Un parmi d'autres* [*One Among Others*] as if to count me, too, his present analyst, among these dead, while alluding at the same time to the number of victims in the other dream.

It would be useful to reflect on the perspective of adventization of the dreams of the analysand in question. His chief problem concerned the sexual impotency that had befallen him following a trivial telephone call he had received while he was in full loving embrace with his future wife. This symptom generated great anxiety in him. The aggressivity contained in his dreams was aimed at adventizing such anxiety, anxiety that accentuated an obsessional neurosis that had been well established since his early youth.

The fundamental mode of the appearance of the subject in dreams: a first approach

Aventisation (adventization) is man *en avent*[6] of himself. This movement forwards (*vers l'avent*) is played out scenically and dramatically. Only the narrative play as intrigue is in a position to put desire *en avent*. If man puts himself *en avent* of his own being, this division or duplication is not, however, a mirror in which he only sees himself. The *avent* is the subjective division of which the agent is an other than myself. This *alterity* is the dialectic of the same and the different. The "I is an other" ("*je est un autre*", Arthur Rimbaud), another who is bound up with history, culture and society and who perpetuates himself through individuals without being reduced to them alone. The dream is the diction of the Other, thanks to which the subject lets himself tell stories and fictions, nourished certainly by his past, but played out from the point of view of his future put in the present. Each dream is like a pastiche that borrows its style from one of the narrative structures belonging to our cultural heritage.

In the dream, the subject "imitates himself" (*mimesis*[7]) as he will have been (*aura été*). But having returned from this future perfect, he reveals himself as the Other. Dream narration constitutes itself as a third party in order to reveal me as Other to myself. Thus it is at once openness and tearing. The narrative web of the dream is, in effect, "a darned stocking" (Hegel). It is the distance that separates me from the Other and which tells me my own history as it would have happened in "a future present". In the dream, we narrate in a "will have been" ("*aura été*") brought back to the intrigue of the present.

It could be objected that the intrigue of dreams is what results from the *secondary revision* of which Freud speaks in chapter VI of *The Interpretation of Dreams.* This means all the subsequent additions and distortions that occur in the dream *narrative.* Admittedly, this objection is not entirely untrue and the analyst is well-placed to take account of it. But one must not lose sight of the fact that speech is not merely a means of communication. *Man has a compelling need to speak about himself and to speak about himself to the Other.* Desire can only be conveyed and set in motion by means of intrigue. Without it, desire would be reduced to rational discourse and devoid of drive momentum.

Whether we participate in our own dream actions or are "neutral" observers of the dream events, and whether the dream narrative is direct or indirect, the narrative mechanism that André Gide calls *mise en abyme* reigns everywhere. "What I like in a work of art," he writes, "is when one finds the very subject of this work transposed to the characters" (Gide, 1939, p. 41, translated for this volume). I am reserving this term for the use which concerns the subject, that is to say the maker of the dream himself, and I am excluding its other numerous applications in literary works.[8] *The mise en abyme of the subject is, consequently, his self-representation in the narrative that is the dream.* It is the appearance beneath the surface of the subject in the intrigue that links him to the Other. In this respect, it is the web on which the dream events are woven.

Certainly, the mise en abyme of the subject is one of the constitutive, if not the most fundamental mechanisms of dreams, but it can also manifest itself as such in a certain number of dreams. Here are a few such dreams.

The dreamer is a 28-year-old man. He works for a company that installs lifts. One day, when he was fitting a control panel, he got electrocuted when he inadvertently drilled through the side of the lift and touched a high voltage cable. He was ejected violently. While he was writhing in pain and before he lost consciousness, he could see the drill which was still turning very close to his face. Since then, the young man has constantly been seeing himself doing the same fateful thing again in his dreams, but by splitting or dividing himself on each side of the panel of the lift. Deafened by the noise of the machine, he wakes up with a start and recalls the words spoken by a colleague who had come to his help at the time of the accident: "You had a very close shave".

All I want to pick up on here is the splitting or division of the young man giving rise to his mise en abyme as subject of the dream. In this example, the mise en abyme is the mechanism whereby the dream attempts to calm down the subject's emotions so as not to interrupt his sleep.[9] It is an abortive attempt because the terror is so intense that the dream cannot continue. It should be noted, however, that the dream attempts to adventize the traumatic event. The trauma is all the greater in that it occurs in a deferred manner after the accident, that is, owing to the importance that the colleague's words assumed. He was said to have had a very close shave (*il aurait été à deux doigts d'y passer*); this past conditional tense in French seems to have been the main purpose of the failed adventization. The failure ends in the recurrence of the dream. I will return later to the question of the role played by the trauma in oneiric activity.

The two following dreams occurred during the same night. The dreamer is a woman who has suffered from two successive failed love affairs. "I dreamt that my ex-companion was working in a supermarket. He was selling fish. I was with my two children and I offered him some plums (*prunes*). There was also a woman at his side who was making very seductive overtures." She then had another dream. "I was with my ex-husband and we were kissing each other very passionately in the middle of a crowd of people. Their presence prevented us, however, from making love. I took him home with me. I was living at the time with two roommates, each of whom was divorced from his wife. One of them was sorely affected by the separation." We can see, first of all, the challenging ambiguity that the dream tries to establish between the phallic attributes of her ex-companion (the fish) and her own children. But the latter "aren't worth a fig" [*ne valent que des prunes*] for their father. It is true that after their separation she had felt devalued as a woman. Thereafter she was constantly deploring her ex-husband's casual attitude and indifference. The dream was a form of revenge on him. Indeed, her ex-husband suffers, in turn, the same setback in the dream through the presence of other men at her home. Beyond her revenge, these men also embody, in the form of a mise en abyme, her own suffering. Likewise, the presence of the seductive woman attests to another configuration of the same mechanism.

In the following dream too we can see multiple mises en abyme of the subject. A man of about forty finally agrees to marry his companion after many years of living together. His companion has long ceased to be the cause of his desire, so he has resumed the practice of onanism to which he was accustomed when he was a bachelor. I am reporting one of his dreams that preceded his marriage. "I was crying over a white kitten that was repeatedly rolling over. With each roll, it was losing a bit more of its vital strength, precipitating it towards death. There was also a woman. She was perhaps about my age or else I was perhaps a child and she was an adult. I can also see a bike . . . and perhaps a railing along the whole length of a narrow street which led up to the house where the kitten was." The latter reminded him of his own kitten when he was a child, probably at the time of the dream ("the bike, the railing and the narrow street"). Two modes of mise en abyme formed around the figure of the "adult" woman. The dreamer is at once the adult that he is in reality and the child that he was at the time when he used to have a kitten. The repeated rolling of the kitten is an allusion to his auto-erotic practices which he hoped to abandon with his approaching marriage, and this was probably the main point that the dream was seeking to adventize. The kitten dies so that "the little boy" can come to maturation. The dream highlights, let it be said in passing, the fear of the disappearance of the subject's desire (aphanisis), caused by his companion's maternal cathexis.

We will end this scenic illustration of the mechanism in question with two other dreams, which are short but very instructive. "I was in a shop that sells women's lingerie. I tried on a bra. All of a sudden, I sensed a masculine presence who *was looking with me as I was looking* at my body." We will move on now to the last example: A woman of over fifty, who is struggling with the fact that she has recently put an end to her alcoholism, had the following dream: "I was in my parents' kitchen. I saw myself as a little girl receiving my father's guests. Yet it was my mother's role to organize this kind of reception. There were a great number of guests. I was in a sumptuous dress. The room was full. I realized that I hadn't prepared anything. Someone had cooked a huge fish. The abundant flesh of the fish was served with two bottles of white wine. The idea that this wouldn't be enough for an evening's partying like that was making me frantic." She then immediately recalled the story of Jesus' miraculous catch of fish in the Gospels, but also his first miracle in which he turns water into wine. In spite of its brevity, the dream turned out to be of an extreme richness that I cannot develop further here. The dreamer had just spent three days "in isolation" with *Marie*, one of her friends who lived in the country. Note the telescoping of her mother and Marie through the intermediary of the theme of feasting and drinking. Note also the current profile of the dreamer who becomes indistinguishable in a remarkable mise en abyme with the little girl that she once was, on the one hand, and with her mother on the other. It is also important to note the ambiguity of the character of Marie, at once the mother of Jesus Christ, who, according to the Gospel story, had told the latter that there was no more wine, and the dreamer's girlfriend who must have sorely missed any sort of alcoholic drink during the weekend in the country. She would have liked the water to be turned magically into wine and, what's more, white wine, her favourite drink.

Projective mode and fundamental mode of the mise en abyme

The device of mise en abyme is at its peak in Proust's (1913–1927) *A la Recherche du Temps Perdu*, where the narrator turns out, at a given moment, to be the one who is preparing to write the events taking place (being narrated) in the novel itself. "In 'autobiographical' narration, one might expect," G. Genette (1972) writes, "to see the narrative lead its hero to the point where the narrator is waiting for him, so that in the end these two hypostases meet up and become indistinguishable" (p. 236, translated for this volume). But the author contests, quite rightly, the possibility of such a fusion. However close they are to each other, the narrator and the hero always have two distinct places. And yet they reflect two modalities of the same subjectivity. Who, then, is writing about whom? "The subject of *La Recherche*," Gérard Genette says, "is 'Marcel becomes a writer,' and not 'Marcel the writer': *La Recherche* remains a *Bildungsroman*, and to see it as a novel of the novelist would be to falsify its intentions and, above all, to force its meaning" (p. 236).

The major characteristic of *La Recherche* is the style or mental mechanism that differentiates what is related from the person who relates it. We often find the same device in oneiric activity. The mise en abyme of the subject does not, however, superimpose the narrator on the one who is striving to be his replica. It should be added that there is no question here of a duplicate. The latter belongs to the order of the specular or the imaginary, whereas in dreams, there is always a gap, a mismatching, which marks the subject with the stamp of the symbolic. In *La Recherche*, this mise en abyme as a division of the subject is practised between the writer (*écrivain*), Marcel Proust, and the narrator/hero, that is to say the one that may be qualified as *writing* (*écrivant*). Likewise, it would be appropriate to speak of the *dreaming subject* (*le rêvant*) insofar as he is different from the dreamer (*le rêveur*).

The subjective mise en abyme that occurs in dreams is neither a superimposition nor a fusion between dreamer and dreaming. It highlights, rather, the *same* as *different*. Which movement can be assigned, then, to the subject who is concerned by the mechanism of mise en abyme? The best formula comes from Marcel Muller (1965): "If the hero coincides with the narrator, it is in the manner of an asymptote: the distance that separates them tends towards zero; it will never cancel itself out" (p. 237).

There is reason to distinguish two modes of the mise en abyme of the subject, the projective mode and the fundamental mode. In the first case, the subject shares the role with one or several characters who appear in a dream. The examples that have just been cited illustrate this. By the fundamental mise en abyme, we should understand the relation of alterity that the dream subject maintains with himself, namely, the relation, as mentioned above, between the dreamer (*le rêveur*) and the dreaming subject (*le rêvant*) in *asymptotic* form. Thanks to this fundamental mechanism, the subject always ex-sists in relation to himself, lurking, as it were, just in the background. This belongs eminently to the symbolic register of the subject. The mise en abyme of the subject stands in for a loss occurring through the intermediary of an *object* which is often drive-based in nature. In other words, we witness the fading of the subject as it is engendered by the object, cause of his desire. It is incumbent upon the analyst to identify it when he is interpreting.

In the rest of this book, I will often be coming back to this subtle mechanism, at the basis of all oneiric activity, and will add other clarifications to it. It is thanks to such a mechanism that, in dreams, we always have a certain awareness of the fact that we are dreaming. This corresponds, in my view, to what the neurophysiologists are trying to refer to when they speak of the state of consciousness specific to dreaming.

The mise en abyme of the subject is a dream mechanism closely linked to the process of adventization. It constitutes its subjective structure. The intensity of the dream experience as well as the sharpness of the dream image, both of which I will be looking into when examining chapter VII of *The Interpretation of Dreams*, depend on it. The mise en abyme of the subject also plays a major role in interpretation. In this respect it is the analyst's principal support.

The fundamental mode of the appearance of the subject in dreams, final instalment

The interpretation in question cannot be reduced to understanding the significance of the dream. It is important here to clear up a misunderstanding. Interpretation in psychoanalysis is aimed, so to speak, neither at the meaning nor the significance of the dream. It is concerned with the place that the dreamer *occupies* in his mise en abyme as a subject. This runs counter to the criticism that has often been levelled at psychoanalysis concerning the famous multitude of meanings and interpretations. "Logic, the motor principle of free association, on which the theory and practice of psychoanalysis depend," writes George Steiner (1989), "are those of an infinite series" (p. 45). In his view, it is this associative chain that makes the analyst's interpretation arbitrary. All the analyst can do, according to him, is interrupt, in the form of interpretation, the analysand's associations. "It is this *contingent*, purely conventional practice of *interruption*," Steiner concludes, "which made Wittgenstein uneasy about the entire psychoanalytic enterprise" (p. 46, my emphasis). This reproach, however, does not concern the practice of the psychoanalyst in the measure that understanding is not his first aim. "The fact that I have said," Lacan writes, "that the effect of interpretation is to isolate in the subject a kernel, a *Kern*, to use Freud's term, of *non-sense*, does not mean that interpretation is in itself nonsense" (1973a, p. 250). Thus interpretation has the task of underlining the *division of the subject* inasmuch as it is inherent to the dream in the form of mise en abyme of the latter. It is precisely this subjective division that determines the ambiguity of dreams and calls for interpretation.

It is thanks to the division of the subject, revealed right at the heart of the narrative, that the analyst's interpretation has the effect of truth – of *his* truth – for the analysand. When one relates the unfolding of the dream and its clinical interpretation not to the analysand but to an external interlocutor, the latter in effect only receives the interpretation in question in the form of explanation and understanding. It goes without saying that the present book cannot avoid such a pitfall either.

A dream is not made to be understood. It is not the signification, but rather the signifiers that count. These are incarnated, thanks to the mise en abyme of the subject, in a certain number of characters present in the dream scene. But, it should be added, these do not only represent the subject. Rather, the representation involved is effected in a negative manner through the intermediary of an object whose loss is imminent in the dream. That is why the dream often seems like a failed venture.

In the dream of *The Wolf Man*, Lacan identifies the mise en abyme of the dreamer not at the level of the wolves, but of their gaze (1973a, p. 251). The object gaze is a drive-related object, detachable as such from the subject. It is an object that can only come back to the subject in a detached form. Earlier on, I called it the object-little-soul (*l'objet petit-âme*), that is, the subject who is almost nothing but his very object. In other words, the subverted subject only reveals himself through his object, which is always lacking in its place. It is precisely this inexorable lack that causes desire.

The wolves, therefore, in the dream of Freud's patient (Freud, 1918 [1914]) do not represent the subject, but his loss. It is not a matter of taking the wolves as representatives of the subject and, from there, of assigning a signification to the dream. Lacan speaks of an irreducible signifier and not of signification. "Interpretation," Lacan (1973a) writes,

> is not open to all meanings. It is not just any interpretation. It is a significant interpretation, one that must not be missed. This does not mean that it is not this signification that is essential to the advent of the subject. What is essential is that he should see, beyond this signification, to what signifier – to what irreducible, traumatic, non-meaning – he is, as a subject, subjected.
> (pp. 250–251)

This irreducibility constitutes the primordial event of the subject. As we shall see later, it is related to the real (*le réel*) insofar as it determines the incessant return of a remarkable encounter, which, as such, is missed each time. Hence the adventization involved in oneiric activity. The dream, then, is the attempt to adventize what *remains in suspense* in the subject. The essence of the dream may thus be said to lie more in the *articulation of its failure* than in the fulfilment of desire. The day residues, at the base of the dream formation, are, in fact, these signifiers that have remained in suspense during the day. Ordinarily, they constitute a coherent and overdetermined whole whose elements, which are closely associated with each other, *evoke* the failures, the defects, or hiatuses of the subject's past history. But, in general, we only retain an isolated part of these memories of the day before, which are the source of the formation of our night dreams. To be more precise, it is not so much the day residues, however closely linked with each other and characterized by their overdetermination they may be, which succeed in forming the dream, but their equivocal nature, their double meaning, their imprecision or their fundamental ambiguity. These are all signs of lack and loss which acquire the force to question the subject, who can only grasp them in the form of repressed elements. Dreaming is the return of these repressed elements of waking life.

Dreams and madness

The asymptotic mise en abyme of the subject in dreams is strangely evocative of the paranoic person's relationship with the paternal imago. The case of President Schreber is a good example of this (see Freud, 1911 [1910]). Lacan (1966c) eloquently summarizes his problem as "the link made palpable, in the double asymptote that unites the delusional ego to the divine other, of their imaginary divergence in space and time to the ideal convergence of their conjunction" (1966c, p. 477). It was in view of such a conjunction that Schreber focused his delirious ideas around his transformation into a woman in order to be impregnated by the divine powers.

What is the difference between dream and delusion? What saves dream activity from a psychotic breakdown? In dream and delusion alike we are at odds with the Other and we execute his orders. Worse still, we adopt the role that he assigns to us and we take his injunctions as things. Not only is our participation in his dramatization total, but we also place our faith in him.

And yet the dream is the opposite of delusion. It is not the expression of an unbearable suffering aided by an imaginary world which attempts, as in psychosis, to fill the void and the abyss left in the place of the symbolic. It is rather the stage on which the conflict between the dreaming subject (*le rêvant*) and the dreamer (*le rêveur*) unfolds as I distinguished them above according to their asymptotic relation. We will see later on that it is thanks to this tussle, to this conflict tending towards the adventization of desire, that the subject is able, paradoxically, to attain the state of rest that is sleep. The latter is the guardian of his relaxation.

If delusion is nourished by the imaginary by taking words for things, this is not the case at all with dreams. The dream image is the "pictorial" (*darstellbar*) expression of the word, whereas in psychosis, the imaginary is in the service of a real (*réel*) which is devoid of the symbolic. By the real I mean that which is the horizon of our being-in-the-world, the extreme frontier of our universe beyond which we meet nothing but hole and abyss. The imaginary universe of dreams serves the symbolic which is the constant temptation of man to give meaning to the world. It is not, however, a projection of meaning that, like delusion, cuts him off from the world. The symbolic makes common cause with the real which circumvents it and constitutes it by assigning it a frontier, a horizon. This frontier is quite simply crossed in madness. The lack of the symbolic is, in effect, at the origin of the psychotic structure, which amounts to saying that the gap between the subject and the Other has ceased to constitute itself as a third term.

Delusion is a "dream" from which one cannot wake up, whereas the dreaming subject is precisely one who never loses contact with waking life or with his reality. From the depths of our sleep, we never lose sight of where we are sleeping or in what circumstances. The posture of the dreamer is always dependent on what surrounds his or her body. Perrine, a 12-year-old girl abandoned by her parents, always sleeps in a curled up position, the same one that she had when she was in a children's home a long time ago. Just think of the heart-rending narrative of the dream in which a father is watching over the coffin of his child. "A father had a dream," Freud (1900) writes, "that his child was standing beside his bed, caught him by the arm, and whispered to him reproachfully: 'Father, don't you see I'm burning?' He woke up and noticed a bright glare of light from the next room" (p. 509).

The only type of dream that is akin to delusion is probably the nightmare insofar as it is a meeting with the real in outline. It is in nightmares that we are on the point of colliding with the walls and frontiers of our world. But where the healthy reaction of the neurotic causes him to wake up with a start, the psychotic person will continue in his sleep and will witness his subjective degradation without any defences. Stéphanie is sometimes woken up by her husband's outbursts of laughter, laughter that often accompanies his nightmares without waking him up. He is a fragile young man who has been in therapy since he had a delirious episode three years ago.

Sébastien, who is six, told me the following dream after asking me if I would make a collection of spiders with him: "I dreamt of a kind snake that was going through a labyrinth with a view to eating a frog at the other end of it. At the exit, it began eating its own eyes." After inviting him to draw his dream, he asked me to write on the sheet of paper the word ATTENTION in black letters. When I questioned him about the kindness of the snake, he replied that its name rhymed with his own.

The calm with which he told me his dream and the kind character that he ascribed to the snake were both incongruous elements in the face of the deadly anxiety of the little boy witnessing his own degradation. The dream seems to function in reverse to adventization. The oddities of the little boy, which are so frequent in psychosis, are all dramatizations of the failure of the dream to represent the division of the subject. The love-object constitutes itself in the psychotic person not as an ever-missing object, that is, the one who constantly gives new impetus to desire, but in the form of an overflow of the presence of imagery. The fundamental mise en abyme no doubt succumbs to the projective mise en abyme before it can be fully instituted. What we see, then, is an abundance of projection onto strange objects which seek to compensate for the missing nature inherent in desire.

Dreams and phantasy

Once our eyes are closed, another gaze opens which makes up for hearing but also for action, a gaze that depicts what we hear and what we put into action. This is the function that Freud calls dream representability (*Darstellbarkeit*), the mechanism whereby words are transformed into images. The narration inherent to dreams then begins, accompanied each time by the intrigue that is specific to it.

If we accept these conditions, it will be necessary to distinguish two major categories, namely, dreams and phantasy. The latter, however, is not dependent on sleep. Nevertheless, it operates everywhere in the life of the subject and regulates not only his psychic life but also his relations with his fellow-men. Freud ascribes a regulatory character to it which structures the perception of the subject. Dreams, on the other hand, participate in subjective life in a preponderant way by making use of our phantasies. The latter participate in dream formation without, however, merging with it. There are even recurrent elements in dreams, which, as we shall see later on, are the direct expression of the subject's basic phantasies. But they must not be confused with dream activity proper.

Two types of phantasy

We are now going to distinguish two types of phantasy. Then we will consider how they differ from dreams. The first type seems to be more in the service of the ego, for it has a semblance of narcissistic infatuation which nonetheless barely masks its aim of repairing an underlying wound. Remember that psychoanalysis distinguishes the narcissistic ego, responsible for the formation of the *imaginary* during the mirror stage, and the subjective "I" which is of a *symbolic* order, and built, essentially, upon the treasure of language. The first is veil and deception giving an illusion of completeness, while the second, bearing the subject's history, is dependent on the Other and marked by his castration. Here are a few examples to illustrate the category of ego phantasies.

A young man suffering from timidity often phantasized about the following scene, which took place in a stadium. During a football match the crowd of supporters suddenly became immobilized on his arrival, even though he was simply crossing the pitch. Ambiguity was at its maximum: did the crowd of supporters go silent so that he would be obliged to face their gaze or was their sudden immobility a sign of their admiration for him? The gaze, that drive-based object of predilection for the subject, is at work in this phantasy in many respects. It is at once the object of avoidance of the timid person, but also what attracts, like a football, the crowd. This kind of inversion is commonplace in phantasies and distinguishes itself from the ambiguity of the dream. The question that arises is why the phantasy in question was not transformed into a dream. The answer is to be found in the narcissistic charge of the fanciful enactment.

A very obsessional woman often saw herself, just before she fell asleep, free-falling over a town. In spite of her terror at the idea of smashing into the ground, she clearly derived enjoyment from the vision. The phantasy seemed to combine at once her imperative need for control, which frightened her when it took on exaggerated proportions, and her auto-erotic self-indulgence which put her in a state of readiness for sleep.

In the following examples, we will see another type of phantasy where what is at stake is more significant. They constitute the central core of the subjective life of the person.

An anorexic adolescent, who was eating a snack, suddenly saw the image of a dissected heart on her plate. Here we have a typical case of a phantasy in an impromptu form or in the form of a flash. The intertwining of food and desire in this phantasy needs no commentary. The "cherished heart" (*coeur chéri*) that the young girl is for her mother faces her with an immense dilemma. The vision of what she is for the maternal Other frightens her to the point of preventing her from eating in the hope of escaping a voracity that is nonetheless constantly threatening to swallow her up. The content of this phantasy is reminiscent of the fairy tale of *Snow White* (*Blanche Neige*) and the cruel avidity of her stepmother demanding to have the young girl's heart. When asked with whom she was sharing her snack, the adolescent spoke of the presence of her friend, called Blanche, who, what's more, is *dark-skinned*. This contrast of colours underlines what is at stake, that is, the eroticized flesh shared by the young girl and the mother as a coveted object. As it is unable to inscribe itself in its status of lack, the oral object in question is unlikely to ensure the subject's division which would condition her real access to her own desire. The privation of food then takes over from the lack as it is constituent of the subject's desire. It is no doubt because this lack is truly lacking that the phantasy is not transformed into a dream entity.

During a discussion an asthmatic man noticed that an incidental image had irrupted in his mind, only to disappear again immediately: his mother was dancing in front of a medieval city. During an analytic session, the man's associations linked the image to two films of Pasolini, Medea and Theorem. The name of his interlocutor in the discussion, which sounded the same as that of the film-maker, must have triggered the phantasy that combined in the most pertinent way the erotic and mystical love of his mother (Theorem) and the latter's unconscious murderous violence towards him (Medea). Indeed, it was this extraordinary ambiguity that the analysis had brought to light concerning his asthmatic attacks. The principal ingredients of psychogenic asthma can be found here. Faced with the mother's murderous ambiguity, the child responds by the *object* breath inasmuch as it is lacking. Clinical experience often reveals the presence of an extraordinary flow of dreams that alternates with periods of crisis. These dreams, which seek to adventize the lack, succumb to a violence characterized by murder, torture, and blood. The attempt to adventize the cruelty of the mother's phantasies results in failure, for the subject remains halfway between the murder that has just failed and the coming-and-going of breath revealing the mother's anxiety. The almost suicidal behaviour of the child often takes over from the phantasy. The enactment of the perilous incident compensates, therefore, for the psychic dramatization of the phantasy.

Fulfilment or satisfaction of a wish?

Freud constantly posed the question concerning the difference between the two types of phantasy that have just been outlined. It is perhaps in this respect that he speaks of conscious and unconscious phantasies without, however, ascribing a topographical[10] difference to them as such. If we look at things more closely, this difference covers the question that concerns us here, namely, of knowing if there are any grounds for distinguishing dreams and phantasy.[11] Freud explains,

> The wishful phantasies revealed by analysis in night-dreams often turn out to be repetitions or modified versions of scenes from infancy. Thus in some cases the façade of the dream directly reveals the dreams actual nucleus, distorted by an admixture of other material.
>
> (1901, p. 667)

Here Freud seems to confirm the thesis according to which dreams use phantasy without being reduced to it. He studies the mechanism of one dream in order to note their similarities as well as their differences:

> Closer investigation of the characteristics of these day-time phantasies shows us how right it is that these formations should bear the same name as we give to the products of our thought during the night – the name, that

> is, of 'dreams'. . . . Like dreams, they are wish-fulfilments; like dreams
> they are based to a great extent on impressions of infantile experiences;
> like dreams, they benefit by a certain degree of relaxation of censorship.
>
> (1900, p. 492)

Freud establishes a correspondence between phantasy and waking dreams. "I am in the habit of describing the element in dream-thoughts which I have in mind as a 'phantasy' [i.e. secondary revision]; I shall perhaps avoid misunderstanding if I mention the 'day-dream' as something analagous to it in waking life" (p. 491). As he proceeds, Freud comes to attribute the same mechanisms of unconscious formation to phantasy as he does to dreams. But he bestows on each of them different conditions and occurrences. He none the less places the emphasis on the secondary revision[12] of phantasy.

Note that with regard to the question of adventization, the relation between phantasy and dreams is posited completely differently. We know now that the main motive of dreams is wish-fulfilment, which takes the path of adventization. The situation presents itself differently where phantasies are concerned. Whether they are conscious or unconscious, they have one thing in common: adventization does not have a primordial place in them. They are perhaps not alien to a certain attempt at adventization, but their nature and function seem to run counter to such an aim. A phantasy does not seek to *fulfil* a wish, but to *satisfy* it. This is the major difference between dreams and phantasy.

As a general rule, Freud seems to link phantasies to the events experienced in childhood, even if he does not make a basic concept of it (pp. 545–546). This conception of the "reality of fantasies" acquired, as we know, greater importance in the study Freud made of the *Wolf Man*. Admittedly, the relation between the reality of the scenes experienced in childhood and phantasies is much more complicated than I am suggesting here (in particular with regard to their temporality and their transformation into screen memories,[13] or their submission to phenomena of "deferred action" (*Nachträglichkeit*)),[14] but it remains true that, for him, it is a phenomenon based on a certain element of external reality. This leads us once again to consider phantasy as a search for satisfaction rather than in terms of wish-fulfilment. Phantasy could be regarded, then, as a lack of fulfilment, that is to say, adventization, an instinctual drive impulse imprisoned in a non-adventized past. In effect, it is more bound up with the past than the future, a past left in suspense or abeyance, a past that cannot gain access to the having-been, that is, as a present experience.

The diverse avatars of phantasy

There are a certain number of dreams that should be called dreams/phantasies in the measure that scarcely any elements of adventization can be found in them. Their aim may be said to be satisfaction or rather the follow-up to the satisfaction that began in the subject's waking life. Think of all the dreams/phantasies in children and adolescents that follow the viewing of a film, a video game, or the reading of a novel, in which the dreamer takes the place of the hero. The characteristic feature of these dreams is the relation to the reality of the day before. "There are some dreams," Freud writes, "which consist merely in the repetition of a day-time phantasy which may perhaps have remained unconscious: such, for instance, as the boy's dream of driving in a war-chariot with the heroes of the Trojan War" (1900, pp. 492–493). It is not difficult to see that dreams/phantasies are simply a consequence of daydreams.

But more frequently the dream borrows a part of a daytime phantasy and applies to it its own laws of adventization. The following dream of a 10-year-old boy gives us an insight into such an entanglement.

"I was at home, upstairs. My dog Pluto had turned into a monster. I went downstairs to see my mother. She was immobilized in the arms of a tyrannosaurus (a creature, half-man, half-monster, in a film watched the day before). I took the pistol "*à pierre*"[15] and killed it. Then I was at school with my mother. As soon as the teacher wanted to leave, I cried. I brought my mother back home before returning to the school."

Here are a few elements of the child's life. *Pierre* is the second component of the father's first name, who immobilizes the mother in the dream like a tyrant (tyrannosaurus). It is precisely the fact that he shows little interest in his son that saddens the child, who, on the other hand, "clings" too much to his mother. This indeed was the reason for the consultation. As for Pluto, he was born, according to the child, before his birth, that is *earlier* (*plutôt*) than him. He is probably his rival, just as his father is.

The first wish of the dream is expressed thus: "I want to eliminate my father in order to have my mother all to myself." But this first optative gives rise to the second: "I would not like to lose my father; I love him as much as my teacher." He accompanies his mother to the house to stay with the teacher whose remoteness produces in him the same effect of sadness as if his mother had departed. What we are seeing, then, is the transfer of his love for his mother on to the person of his teacher. In this way, the child tries to adventize his Oedipal conflict in the measure that it illustrates the beginnings of an identification with the father. Thanks to the elements of the film watched the day before, the phantasy is integrated with the fabric of the night dream. We notice the gap that separates the father and the teacher. The latter represents, owing to the attention that he gives to the child, what the father is lacking. But it is an imaginary, and not a symbolic lack. So the dream does not possess a sufficient symbolic charge to be able to draw on a lack of which the father is the name (Lacan).

Children's nightmares often reveal their phantasized relationship with their parents. The child's speech is hindered by too much real proximity between the parents and the characters of his nightmares. This is no doubt why the mere fact of speaking about the nightmare to someone else, in this case the analyst, might suffice for the child to free himself from it. It is nonetheless true that the nightmare is an attempt to adventize speech addressed to the parents, speech that tends to be silenced insofar as it belongs to the real (*réel*), that is, to the very constitution of the child's universe.

The recurrence of the same places in the dream is another example of how phantasy is employed in dreams. Clinically speaking, it often indicates that a fundamental individual phantasy has been constituted around these places.

A woman of about forty-five frequently finds that her holiday house in Auvergne appears in her dreams. It was in this house that her grandmother had one day caught her husband having sex with his office employee. This scene, which is also evoked recurrently when she is engaged in sexual relations, shows the place that she would have liked to occupy in the irresistible desire of the Other, a grandfather who was the pleasure-seeker (*jouisseur*) par excellence of the family. I am using the expression pleasure-seeker in the Don Juanesque sense of the term. The legend of Don Juan may be considered as essentially a feminine phantasy insofar as it enacts the quest of Woman. In the phantasy in question, my analysand enacts the wish to be the Woman as an absolute object of desire. Owing to the intensity of the desire, the phantasy had managed to find a path towards the royal road to the unconscious that is the dream (Freud). In effect, it enacted the other side of her subjection to the desire of the Other, that is, to be violently desired rather than humiliated. The phantasy about the grandfather was designed to repair a narcissistic wound dating back to her first love in youth. By falling madly in love with one of her school friends, she had finally succeeded in making herself desired. When it came to the moment of making love, the young boy had got up and suddenly left the young girl who was now ready to give herself to him. We can imagine the young man's fright when faced with the girl's ardent desire, something he was discovering for the first time. However, this experience, she said, had left her feeling irremediably degraded. This failure had forever offended her sensibilities. This is why she constantly invoked it in an inverted form in her phantasy.

Here is the recurrent spatial element in the dreams of another patient. It involves picturesque landscapes viewed through a window frame or a chassis frame or through binoculars, landscapes that are fixed for a moment in order to be admired. This also involves a phantasy. It highlights the patient's problem concerning the spectacle of seeing the naked body of his blind grandmother when he was 7 years old. The sight was all the more exciting in that it took place in the presence of the *gaze* of a blind person, one that has become a gaze par excellence owing to its inherent lack.

It is easy to see why Lacan says that the end of analysis is the crossing of one's own phantasy. The phantasies in question are loaded with such *jouissance* (overflow enjoyment) that the subject refuses to lose them, but also to deliver them up to the knowledge of the Other. This would explain why phantasy is the failure of adventization. This failure is consubstantial with phantasy; hence its recurrent character.

The failure of adventization in the phantasy is closely bound up with its temporality. This resides in the haste in capturing what might be called the *liminal point of jouissance,* just at the moment when desire reaches the height of satisfaction before disappearing once again. Haste tries to capture the ephemeral instant of *jouissance.* The image of the enjoyable scene is immobilized in haste, haste that has the function of avoiding, that is to say of concealing the imminent loss of desire (*aphanisis*). The subject tries in this way to tear the "real" (*réel*) away suddenly from the *jouissance* contained in the scene before its evanescence.

The liminal image is a fleeting glance at the scene of *jouissance*. The simultaneity of the image confers on it a synchronic temporality that runs counter to the diachrony of speech. The image contained in the phantasy partakes of the configuration of the subject's history, but in the form of an enjoyable fixation. *Jouissance* should be understood in the sense of the inflation of pleasure, of a "more-than-enjoying" ("*plus-que-jouir*") (Lacan).

In the phantasy, we are bogged down in the image as a hasty capture of a moment of *jouissance* whose liminal point, which has barely appeared, once again slips between our fingers. *Phantasy is, in effect, a defective mode of adventization.* To be more exact, it is the concealment of the time of desire. In the phantasy concerning the pleasure-seeking grandfather, the statement, "You will never be the Woman, the absolute object of desire" is concealed, like a barely audible true saying, in favour of the image of the one who abandons herself to the gaze of the man whose *jouissance* is supposed to be limitless. The instant of the grandfather's gaze determines henceforth the liminal point at which the imaginary phallus, in the form of the seeing/being looked at of the phantasy, is exchanged between the subject and the pleasure-seeking Other. This absolute object of desire has the function of blocking out the lack in which the subject constitutes himself as divided.

Imaginary dreams

There are dreams formed from phantasies that might be qualified as ready-to-wear. They are forged principally on the basis of an ego construction, that is to say, they are purely imaginary. The extraordinary strength of imagination that they deploy should not lead us into error: they remain devoid of the symbolic depth that we are generally familiar with regarding dream formation. We may call them imaginary dreams to distinguish them from those that belong to the symbolic register. Their imaginary dimension does not call for interpretation, but explanation. Their narrativity often stems from stories seen or heard but which are caught in the web of the dream logic and reorganized to this effect. This logic nonetheless remains weak in them. They are more akin to reverie and dependent on daydreams. The question that arises is that of knowing their aim. Is their task to give more consistency to the construction of the ego, that is, to bolster the dreamer's infatuation and self-sufficiency, or is their purpose also to adventize a trauma, a narcissistic wound, or a breach of self-esteem? The following case concerns a positive ego construction in which we can discern in the background a limited attempt at adventization thanks to the progress of the analysis and to the support from which the subject benefitted during the clinical work.

Sofiane is an 11-year-old boy, who is slightly mentally retarded. His problems at school were such that he was soon oriented at the age of 9 towards a Medical Educational Institute[16] where he was finally able to have a place.

This orientation inflicted a narcissistic injury on him, though. He constantly felt the need to compare himself with the other children he knew from his area who continued to attend the school. The analytic work enabled him to rebuild his self-image. His many dreams revealed his attraction for the world of champions of football with whom he identified. His problems of intellectual aptitude should not mislead us. Boys at the age of latency very frequently have such dreams and they are often bound up with their play activity. Later on, this kind of imaginary construction will merge with the problematic issues of adolescence and give rise to a different mode of dream representation. Below we will find an illustration of this in the dream of a young girl of fifteen.

"Someone had discovered or invented life after death. It was a man aged 45–50, with long and stiff hair that left his forehead bare. He had killed a black man who had refused to obey him. In spite of that he was not wicked, but friendly. I was with a group of young people. We were looking for the door that was supposed to lead to the place where there was life after death. We were inside a grey building. After some searching, we got to the place where the door was. I cleared the access to it by moving two or three cupboards that were blocking it. Finally, there we were in front of the door; but I was very apprehensive about opening it. My companions were encouraging me to go ahead. Finally it opened. Behind it there was a vast verdant landscape with hills; in the distance, an old man was bent forward over his brown walking stick. Near him there was a group of young people, who were in fact "living" dead in this place supposed to be the place after death. Among them, I could see a young girl with a fluorescent yellow pullover."

She brought few associations and those she brought were imprecise, contributing little to our understanding of the dream. The inventor of life after death made her think vaguely of the former tennis champion, Henri Leconte. The young girl had seen him recently in the programme *La ferme des célébrités* [The Farm of Celebrities]. We were unable, however, to link it up in any way with the dream content. Similarly, the young girl in a fluorescent yellow pullover made her think of a classmate, but without any further links. Nevertheless, one thing became a bit clearer to her: the verdant landscape seemed doubtful to her, that is, it was mixed with anxiety.

When asked about her recent state of mind, the dreamer was unable to identify any particular states either of joy or sadness. "It's the usual daily round of life," she said, "without anything particular to report." The dream seems as enigmatic as the question of life after death.

Without sufficient associations, the possibility of interpreting of the dream seemed compromised. So I will venture an explanation rather than an interpretation.

The beginning of the dream seems to be an act of initiation, a group of young people setting out to look for life after death, discovered by a doubtful master. Initiation is, in fact, a juvenile quest of prime importance. The ambiguous and doubtful character of the master probably belongs to the process of initiation. Moreover, it was no doubt in this connection that he had killed a disobedient disciple.

We know how young adolescents are not only extremely concerned about their image and self-esteem, but also about questions relating to their future, which sometimes take on an existential dimension. Furthermore, we know how important the register of the imaginary world and the ego is for them.

The main question for the young girl is to know what would become of her if she were to put an end to her life. The dramatic aspect of such a question is, however, deceptive. In the uncertain image that the young person has of him/herself, in his constant quest for identity and in his/her ongoing search for new identifications which is marked by lability and change, everything proves possible, including the eventuality of death. Such thoughts or preoccupations can either take on the form of ego mastery or put themselves in the service of a certain depersonalization[17] without any grave consequences. It is with reference to such psychic coordinates that we can approach the meaning of the dream.

In dreams in which phantasy, the imaginary and mastery prevail, associations are often lacking. It would doubtless be pointless to look in this kind of dream for a desiring dimension in the unconscious sense of the term. To reduce all forms of dream activity to the formation of the unconscious would therefore be an error.

The waking residues sometimes seem to play, as I have said, a preponderant role in the formation of imaginary dreams. This is the case of dreams formed after watching a film or listening to a striking story which depicts the raw and brutal aspects of the partial drives. It is precisely in this respect that these waking residues elicit the specular image in the subject and animate his partial drives.

"I dreamt that I was living on a desert island with my family. At the same time, the island resembled a spaceship stationed on the sea. Our house was all in glass. It was my birthday. I was getting ready to blow the candles out. Suddenly, I saw a bizarre object appear right in the middle of them. I took hold of it without fearing the flames. It was a strange speaking clock standing on its two legs. Then it was pressing against my chest. I realized with fright that it was a baby Alien. I woke up with relief."

The dreamer is a fan of horror films. Most of the elements of his dream are in effect borrowed from cinematographic productions. The spaceship stationed on the sea is a combination of two films of the same series. The first, *Alien, the Eighth Passenger*, involves a spaceship called Nostroma which has to make a forced landing on a deserted planet; the second *Alien, the Return*, tells the story of a volcano (note the flames in the dream) on a Pacific island where two journalists make terrifying discoveries. On the other hand, none of his associations explained the speaking clock. But the fact that it speaks and is equipped with two legs gives it an eminently uncanny aspect.

This uncanny ambiguity is, generally speaking, an integral part of horror films, which may be distinguished from nightmares proper.[18] It is linked to a constant and constitutive element that can be found in all imaginary dreams, namely, the duplication of the ego, when the ego finds itself face to face with itself and when symbolic alterity (i.e. the subject's division) proves to be absent. In the dream reported above, the duplication is invoked in the form of the relation that the dream establishes between the birthday and the baby alien.

The cinematographic creature, Alien, shows very eloquently the duplication of the ego insofar as it appears after the narcissistic mirror has been traversed by the subject. Once the curtain has been lifted, we can see behind the lure of the narcissistic image that uncanny double of ourselves which knows nothing but ugliness and terror. The specular image that is formed at the mirror stage takes place only in aggressivity and auto-eroticism. This, moreover, is why sex and violence are two major ingredients of horror films.

One of the characteristics of horror stories – nourishing so easily the dreams that we are studying here – is the sudden appearance of an evil being just at the moment when the subject is abandoned to the solitude of his ego. This obviously does not occur when he is in the presence of the symbolic coordinates. It is simply that the solitude in question is on the lookout for the evil that is within him. *Alien-Clone*, or *Dracula* fleeing his specular image, are both forms in which the duplication of the ego appears. The latter may be considered as the failure of the dream mechanism that is the mise en abyme of the subject. The double as a product of the narcissistic ego is what manifests itself in the absence of the dialectic inherent in subjective division. Indeed, it is such a division that makes the mise en abyme of the oneiric subject possible. Without such a dialectic, the division yields to the abyss of the uncanny relationship that the ego maintains with itself. This relation is governed by fascination.[19] This is all the more accentuated in that the ego remains stunned by its own horror. The more it is captivated by the narcissistic image, the more it stares at it, and the more it derives enjoyment from its own violence and auto-eroticism. It is precisely this fascination that drives individuals to watch horror films. It is doubtless the same abyssal fascination that presides over their dream production.

The dream may begin with an adventizing aim, but it soon falls under the spell of the fascination exerted by the captivating image of the fragmented body. This may explain the reason for its occurrence. Henceforth, the partial drives, which are at the origin of dream formation, put themselves entirely in the service of the specular ego, which is dominated by aggressivity and auto-eroticism. In horror films, it is rarely a matter of putting on screen sexuality in the subjective or even pornographic sense of the term; it is simply auto-eroticism projected onto others in a relationship marked by depersonalization.

In dreams of horror, the ego is unbridled and that is why it cathects the partial drives so intensely. The scale of the instinctual drive pleasure/fright is then such that the only outcome and the sole way out of narcissistic horror will be waking up with a start to rediscover the intact image of the non-fragmented body.

Trauma and its consequences

Thanks to the notion of adventization we are now in a position to understand a certain number of subjective phenomena that were not easily understandable before. Dreams relating to traumatic events belong to these phenomena. As a general rule, they may be divided into two groups: iterative and singulative.

A. The iterative occurrence[20]

It goes without saying that the occurrence of these iterative dreams is governed by the attempt, that is to say, the desire, of the subject to adventize the traumatic event. They indicate that on each occasion the attempt is aborted. The traumatism is constituted insofar as the event concerned cannot, owing to its scale, its force, and its intolerable content, be integrated with subjective life. In other words we are dealing with an entity that threatens the subject's integrity.

Broadly speaking, there are three key moments in the Freudian elaborations concerning the question of trauma. The genesis of hysteria was first conceived by Freud as the consequence of a trauma, in particular, a sexual trauma, during childhood. This conception goes back to the period prior to 1900. At that time the phenomenon of seduction was at the centre of his theory. Thereafter it lost the dimension of reality that Freud had attributed to it and came to be considered as phantasy. The so-called theory of seduction, which Freud refers to as his *neurotica* in his letters to his friend Fliess (letter dated Sept 21, 1897 in Freud, 1985) was supposed to explain hysterical symptoms in terms of a traumatic experience of sexual seduction that allegedly occurred in the patient's childhood. It was not without a certain disappointment that Freud finally understood that this was not the case. The seduction at issue in the narratives of patients had no other origin than their own phantasies. Thereafter, it was often counted among the *primal phantasies* such as castration or the primal scene (imaginary reconstitution of the parents' sexual relationship) to which Freud was inclined to ascribe a phylogenetic status, inherited from our ancestors.

Beyond the Pleasure Principle (1920) traced the question of trauma back to the compulsion to repeat, and consequently, to the death drive. There is indeed an intrinsic relationship between the persistence of the trauma, notably, in the form of dreams, and the failure of adventization concerning it, hence their constant repetition.

In *Inhibition, Symptoms and Anxiety* (1925b), Freud conceives the occurrence of iterative dreams of the trauma as signal anxiety caused by the possible reemergence of the event. The attempt at adventization as a constant quest of the subject to come to terms subjectively with the impact of the trauma seems to me to be close to the Freudian conception of repetition as signal anxiety. The dream tends to provoke the traumatic event with the aim of remedying it through adventization. The subject thus tries to consider the trauma from the point of view of a future from where the event is apprehended as belonging to the past, as having-been. The trauma is in fact a psychic phenomenon that is incapable of transforming the past into the having-been of the subject. It is the perpetual present of a past that is constantly returning to the same place, whereas adventization is the reverse process.

A chaste woman, over 50 years of age, often has dreams relating to the same recurrent theme, the cold. These are not iterative dreams, strictly speaking, but dreams that constitute either variants of the same trauma or are linked directly or indirectly to it. And so it seems sensible to classify them under the same banner.

Let us begin with one of her countless dreams during her analysis. "You and I were in a cellar moving around the freshly turned over soil. Your eyes met mine. I suddenly felt afraid, even though I normally sense a lot of warmth in your gestures and attitudes. In the cellar, to our right, there was a bright skylight contrasting with the darkness of the cellar. On the other side of it, I noticed a little girl sitting on a stack of wood logs. At that point my mother arrived; you weren't there any longer. I was looking out of the window on the first floor at my parents' house. I could see my maternal great uncle felling trees in the distance." At first sight it is a dream without intrigue, except for the exchange of looks, an allusion to the couch where the gaze ceases in favour of speech. The gaze as lack causes, therefore, the desire of which the subject is merely the effect. This reminded her of the gaze "full of voluptuousness" that her depressed father cast at her in those rare moments when he emerged from his chronic state of melancholy. Indeed, the father's apathetic state stood in contrast with his indecent looks which petrified his daughter and gave her a feeling of being cold. "You know, the sort of sensation that invades you when you are in contact with soil." This took her back to the burial of her maternal grandmother, when she was laid out in the coffin. This grandmother was without any doubt the only light in her childhood, the only person with a healthy soul who was able to give her love, but also protection.

Her mother experienced moments of intense agitation during which her body was capable of making big starts or jumps in the direction of the ceiling while she was lying in bed. This mother, whose intentions were ambiguous, was capable, for example, when she was preparing a chicken for dinner, of quite simply saying to her daughter that one day she would inflict the same thing on her. She refused to have sexual relations with her husband, but would go on the slightest pretext to see her uncle, a woodcutter, who most probably abused her sexually. This sexual ardour stood in contrast with the coldness she showed towards her husband.

For our dreamer, the cold obviously stood in a dialectical relation with fire and warmth. The logs on which the little girl was sitting, in the form of a mise en abyme, and which were destined to be consumed in the fireplace, speak volumes about this. Felling trees is a euphemism for the question of castration linked to the person of the great uncle, the woodcutter. Not only was he club-footed, but he also cut down trees. What else was her mother trying to do if not to castrate this man who had always remained a bachelor? Being subjected to humiliation by him was worth it when set against the castration that she inflicted on him; this at least was the point of view of the little girl concerning these incestuous events that were going on around her.

Her mother cooked up a pretext one day to send her round to the neighbour. No sooner had she arrived than she was made to lie down on the bed before being immobilized under his enormous weight. She was paralyzed with fear and could only implore God to come to her aid. At this point, the neighbour's son came into the house and so the man was obliged to put an end to his dirty work.

Barely aged ten, the little girl returned home in a state of total disarray, shivering from cold for several hours. No one bothered to ask her why she was so distressed. This was how she experienced the trauma that would determine her future life. The trauma turned out to be even more devastating in that it was linked to the incestuous desire of her father who was constantly trying to kiss her on the mouth or to wedge her in the corner of the living room.

"My father appeared in the doorway with a gun in one hand and a rabbit in the other. Blood was still dripping out of the animal. My mother held out a parcel for me. I took off the wrapping and inadvertently put it on the gas cooker. The fire spread through the kitchen. My father took me in his arms and carried me to the neighbour's where there was apparently a party. I woke up and asked myself if I had turned the gas off properly in the kitchen. I was so obsessed by the idea that I went to check." In this dream, as in her trauma, the object, cause of desire, is situated at the junction point between fire and blood. This is where her subjective division lies, between hot and cold, between her frequent amenorrhoea and her recurrent sensation of menstrual blood flow, between the mortifying mother and the sadistic father, between the appearance "all fire and flame" of the latter in the doorway and the sending of the little girl to the perverse neighbour's house to have her "worked over".

In the course of her analytic work, she would often repeat the same sentence: "That man who assaulted me, left me bloodless." She did in fact have blood losses without any apparent reason. The analysis established that these losses occurred when she had contact with men whom she particularly liked. It was not uncommon for her to find herself speechless and paralyzed, as it were, in their presence. Her body, which flared up in response to the looks they gave her, reminded her of the incestuous flame that linked her to her father.

"I dreamt that I was still living with my parents. We were in the process of moving house. In the lorry, the driver, who had a moustache, put his hand between my legs. I was unable to move. The man took out a rosary and put it around my neck. It was the Hand of Fatima engraved on copper." The ambiguity of the hand is clear to see: was it a case of aggression or of protection? The driver with a moustache is a composite figure. The man who had assaulted her was a lorry driver. But the moustache reminded her of her maternal great uncle, her mother's lover. At the time she was keen to move so as to avoid having every day to face the malicious looks of the neighbour who had assaulted her. As for the copper on which the Hand of Fatima was engraved, it reminded her of the entrance gate at her analyst's where she came in search of refuge and protection, but also the erotic love evoking her assailant. Such ambiguity often stems from the disaster caused by the trauma.

B. The singulative occurrence[21]

The second type of dream, concerning traumatic events, could be described as singulative. As their name indicates, these dreams do not comprise a repetitive aspect. They occur during analysis with a view to being adventized. The particular arrangements of analysis will enable the analysand to allow himself to speak about them.

The analysis will therefore have a liberating effect so that repression, pertaining to the primary process,[22] ceases in favour of a more integrative mechanism. In the absence of repression, we can see the process that had led the subject to adopt a low profile to avoid admitting to him/herself the traumatic character of the event. And so we are sometimes astonished to see that patients remain convinced that they have already talked to us about their traumatic event. It is not unusual for them to go as far as to deny its devastating character. This might even lead them to operate a sort of split between the part concerned by the trauma and the rest of their body.

A female analysand of about fifty suffered from insomnia and symptoms bordering on psychosis. She insisted firmly that her terrible disarray at night had nothing to do with the incest that she had been subjected to for many years at bedtime. As proof of this she cited her "normal" sexual life with her partners. According to her, only the lower part of her body had been concerned during the fateful act. But now it was her whole painful existence that was suffering from it. Here is another example.

A woman approaching forty had the following dream: "During a marriage, I noticed that the bride was paralyzed." Another dream followed in the same night: "I was with a boy whom I had known at the age of 18. Our attempts to have a sexual relationship at the time were unfruitful. I dreamt that I was with him, but in the presence of a rather unpleasant man who was unknown to me: all three of us were getting ready to make love."

The paralyzed bride is an allusion to the symptom of the dreamer herself, a partial and intermittent paresthesia of the right leg with sensations of liquid flowing on the inside. As for the second dream, there were no associations to help us. When I asked her if there could be a link between the two dreams, strangely enough she replied by mentioning her recollection of a traumatic event. She was sure that she had already spoken about it. We understood subsequently that her apparent certainty was part of a set of defensive strategies that she had adopted to ward off the suffering due to the trauma.

One evening, on leaving her boyfriend's appartment, she had the feeling someone was following her. She was only nineteen at the time. Night was falling and the streetlights were not yet on. She turned around suddenly and shouted at the person following her: "Stop following me!" "Calm down, *Mademoiselle,*" he replied, while grabbing her from behind. She said that he had "one hand on my right thigh, the other on my left shoulder." The part of her body referred to corresponds, in fact, to the zone of paresthesia mentioned above. "I got my umbrella out and didn't hesitate to hit him. As soon as I managed to get free of him, I started running. A group of young people then barred my way, but I slipped down a side alley and escaped. It seemed to take an eternity to get to my parents' home." When she finally arrived at the house, she noticed her periods had started. This was followed by a sleepless night during which she couldn't stop brooding over the scene of the aggression.

The second dream concerning the unfruitful love with her boyfriend in the past seems unambiguous. It alludes to the traumatic event, that is, to the attempted rape, but in an inverted form: "If only I had met this young boy from the past instead of the aggressor!"

It is part of the analyst's clinical responsibility not to prejudge such ambiguities concerning the dignity of the victim of the aggression. This dignity divides the subject. And the process of adventization attempts to transform it into desire. There is nonetheless a remainder that scarcely lends itself to this kind of symbolic treatment. The veritable point of the traumatism in the victim is in fact this remainder that has the unfortunate tendency to awaken the old demon of *jouissance*[23] (overflow enjoyment) that is dormant in each one of us and ready to surge forth at any moment. That is why *jouissance* is nothing other than suffering. To take *jouissance* for a form of desire would be a fundamental misunderstanding. It is true that the *jouissance* in question has all the appearance of desire due to the ambiguity of subjective division, but also to the attempt at adventization. It is to the credit of the clinician if he or she can avoid such confusions.

Our analysand's traumatic experience had struck precisely the area of the body that was already fragile. It was indeed this same part of the body of her mother that she had seen accidentally at the age of 14–15, a troubling sight engendering a subjective division in the form of erotic excitation that was immediately repressed. The paresthesia of the zone concerned on her own body seemed to come from the conjunction between the aggression and the sight of her mother's denuded body. Here we have an "overdetermined element of chance" of the kind that I will be discussing later on, and which brings together two essential elements for the constitution of the reality of the trauma, namely, the chance element of an encounter (*autómaton*) and its resonance in us which is fortune (*tuché*).[24]

To this may be added her father's favourite game. In the evenings, as she was about to go to her bedroom at the end of the corridor, he would often approach her from behind and mischievously try to frighten her. Significantly, following her recollection of the assault, she said that it was as if all the scattered elements of her life (*autómaton*) had come together there and suddenly acquired meaning (*tuché*, fortune). "Before," she said, "the assault only had an anecdotal significance for me. It's now that I am taking it fully into consideration." Hence her feeling that she had already talked about in the analysis.

The sight of her mother's body as well as the exciting game with her father were fully taken into account during the analysis. This was not, however, the case, for the assault; and yet it was not a repressed memory. The analysis was able to show that the two events, both sources of excitation (the sight of her mother's body as well as the game with her father), had so to speak, "made a fortune" (*fait fortune*), that is to say, they had acquired increased erotic meaning retrospectively (*après-coup*). The aim of this was to adventize the assault by bestowing on it a playful/sensual meaning. The assault was supposed, therefore, to become a source of excitation instead of gnawing away at her mind. In other words, the *jouissance* was paradoxically supposed to act as a remedy for her distress, even though it was a source of suffering passed on to the body.

When the assault took place, she had felt the weight of a profound solitude – the solitude, precisely, in which her mother had always left her. She had lacked warmth from her mother, who was sick and blinded by the attention that she devoted to her eldest son. "When the assailant let me go, I looked at the buildings; none of the windows were open. I saw the façades of buildings but there was absolutely nothing and no one to help me. I didn't stop yelling, though. I ran and ran, and felt unreal. I was a spectator." Faced with this unreality, with this division provoked by the encounter with the real, the body takes over. It enacts what could have happened, like a missed encounter. This, in fact, is where the real (*réel*) is reconstituted, that promoter par excellence of *jouissance*/suffering due to the chance of an encounter (*autómaton*) that ultimately brings fortune (*tuché*), that is, by awaking the demon that is lurking in us. This would explain the occurrence of periods. "During the whole night that followed the assault, I thought about it. I woke up feeling very stiff, as if I had been running all night." So the body repeated the assault. Repetition is, in effect, the major characteristic of the real, which always comes back to the same place (Lacan). The appearance in dreams of the traumatic event resembles a theatrical rehearsal whose temporal vector is inverted. It is the same theatrical enactment that governs the process of adventization, except for the fact that the latter espouses the expectation of the subject through the grammatical intermediary of an accomplished form.

* * *

There are dreams in which the subject manages to anticipate the occurrence in the immediate future of a traumatic event. Here is an example. A woman, very advanced in years, related a dream from her early childhood. Her mother had been very ill since her birth; it was considered that her suffering was such that she was not likely to survive. But now the symptoms began to get worse and the family was trying to do more in the way of helping her. It was then that she had the dream that had been engraved forever in her memory since she was a child of barely 6 years of age.

"I had left home by myself and was making my way to school. I was carrying a sort of basket on my head, like African women. Suddenly, I stumbled and the basket fell to the ground; it sank deeper and deeper into it. I tried desperately to unearth it. It was a lost cause; the more I tried to dig it out, the more I couldn't find the basket and the more I was assailed by anxiety. My sister, who was 8 years older than me, then appeared at my side. She told me that it wasn't worth continuing; the basket could not be found. Then she helped me to get up and took me in her arms."

We don't know if a screen-memory is involved. We can also discard the possibility of ascribing a premonitory character to the dream. It is easy to assume that the dream was merely an attempt to adventize the imminent death of the mother who was suffering from an incurable disease. The dream spoke about what the family refrained from revealing to the child. She thus tried to adventize this tragic demise to the benefit of her elder sister, who was invested as a maternal substitute.

Figures and numbers in dreams

Among the major signifiers we can speak of figures and numbers. Their relation to death is so frequent in dreams that one is tempted to make a general principle of it. As Freud taught us, dreams are often formed on the model of linguistic devices, and, in particular, tropes and figures of speech. In this connection, one would not be mistaken in considering the figure and its occurrence in dreams as figures of speech that are related more specifically to the question of death. We would thus be justified in calling them *figures of finiteness.*

Freud devoted a whole section to numbers and calculations in *The Interpretation of Dreams* (1900), though without establishing the relationship we are concerned with here. And yet the examples of dreams that he gives very much confirm my hypothesis. We shall see that his relationship with Fliess, whose passion for numerology he was well aware of, was directly related to it. Here is an initial example of the relationship of dreams to rhetorical figures and the ultimate destiny of human beings.

Someone who had suffered throughout his childhood from a lack of paternal love, rather unexpectedly took care of his wife's stepfather, a distinguished man, well-advanced in age, and almost disabled. Something the stepfather had said to him had left him feeling literally stunned: "You will be my son!" Thereupon he had the following dream: "Everyone was there, my parents, my sister, my wife and her family. We were in a village. An old man with a walking stick, whom I was caring for, was living with us. I went out in his company. I had to take him somewhere. The taxi stopped near a pick-up point. A lot of people were waiting. Further on, there was another car with a couple inside who asked me if I had any minors. I said no. They said it was only an eleven-hour-drive to Ivry."

To comment on the dream, I will begin at the end. *Ivry* is to be understood pho-netically, that is, as relating to the drunkard, the alcoholic (*ivrogne* in French).[25] In fact, the evening before, our dreamer had watched a television programme on resilience. There was mention, in particular, of a family of six children who had lived in the hovels with an alcoholic father, but who had nonetheless managed to pull through. The question raised in the dream was: will this old man, who wants to occupy the place of a loving father for me, be resistant enough to pull through? Or will he die shortly, "in only eleven hours"? This question of rehabilitating the father is rather ambivalent (cf. "Do you have any minors [*mineurs*]?"). The signi-fier *mineur* had a double meaning for him: someone who lays bombs or mines (a sapper), but also the contrary of an adult person (*personne majeure*). It is true, he added, that only "minors" are "capable of being adopted". If one only took the aggressivity towards this man into account, it would be reductive.

True, the number eleven evoked for him the expression, "*boire le bouillon de onze heures*" (lit. "drinking the 11'o'clock broth", i.e. the poisoned broth), which is in fact a death wish. The reason for this aggressivity lies rather in Freud's remarkable discovery concerning the primal phantasy of the murder of the father, a psychic disposition of foundational significance for the advent of the *symbolic* place of the father (see Freud, *Totem and Taboo*, 1912–13). Here as elsewhere, the symbolic death of the father is closely bound up with the birth, real or metaphorical, of the child. The number eleven corresponds in fact to the dreamer's day of birth. Here we are at the peak of the mechanism of adventization in the oneiric impulse. The dream seems to adumbrate a form of adventization concerning paternity. Birth is the metonymy of death.

Our dreamer bears the maiden name of his paternal grandmother. As he was not recognized by his biological father, his father had no choice but to take the family name of his mother, who was herself born of an unknown father. The dreamer had suffered during his childhood from a lack of affection from a father who had difficulty assuming his role. The only aspect of paternal affection that he recalled was its educative dimension. Like an ineffable wound, this deficiency still continued to pursue our dreamer. He obstinately refused to respond to his wife's wish to have a child. However, the question of adoption became all the more acute in that it now rested on the affection of a man (the old man in the dream) grappling with a truly paternal love for him. The number eleven, indicating the day of his birth, expressed this clearly. Will it be a question for me of a new – but real – father? Would making his adoptive father's love an offering to the other be strongly charged with affection? But this aroused all the aggressivity that he had never allowed himself to express towards his own father. The old man, whose intentions were nonetheless innocent, could not be spared this ambivalence.

The most crucial question for him now was this: how can I have a child and bequeath to him what I have not really received from my own father? His family name attested to a lineage that was solely maternal. Neither on the side of the old man who had offered to adopt him, nor on the side of his wife enjoining him to become a father, was any form of subjective division possible in order to give rise to a truly symbolic dimension. Paternity seemed to him more like a pick-up point where one cannot really choose to live. And yet the symbolic elements of the consecration were united; the whole family was clearly convened to this end.

Below is a further example of a number that functions as a figure of finiteness.

"I dreamt that my son was ten and a half years old, but that he had been dead for three years. It was unbearable and I didn't want to know anything about it; I had a writing project . . .". Ten and a half is, in fact, the real age of her son. The "three years" is an allusion to her separation from the son's father. The death that occurs in the dream is the same experience of death that the patient is going through in her mourning with a view to adventizing the trauma of separation. This unbearable projection reveals its true nature at the end of the dream where there is a reference to a writing project, a work that would survive her own death. Here is another example of numbers in dreams.

Melanie consulted me about her anxiety at school related to the possibility of failing. She is 10 years old. Her twin brother had to repeat the class he was in last year. During the first session she was accompanied by her father, an evasive man, who had no doubt been "sent" by the mother who did not want to attend the session. The question of the fusion between brother and sister was touched on in the context of the family history, though nothing particularly significant emerged. The individual interview with Melanie nonetheless proved fruitful. In the next session, she told me about a nightmare she had had: "I was alone in our village, it was completely deserted, and then I woke up with a start." Here we are no doubt witnessing the beginnings of an attempt to adventize her "separation" from her twin brother. She then wanted to do a drawing. She wanted to draw, she said, "*five* children" (the actual number of her sibling group) and, making a slip of the tongue, "*two* fathers". Her father does, in fact, have a twin whose name begins with the same initial as his. In Melanie's drawing, however, there are only four children instead of *five*. She had forgotten her brother.

The little girl feels absolutely alone insofar as her father is unable to help her find a solution to her fusion with her brother. The father most probably has the same difficulty. The mother's attitude of withdrawal increases this solitude. One wonders how they could even take the step of seeking a consultation for their daughter.

The singularity of this case resides in the extreme affinity that is noticeable between the drawing (forgotten brother) and the content of the dream highlighting the solitude feared by the little girl, a solitude experienced in the form of death-anxiety ("I was alone in my village; it was completely deserted"). It is through her twin brother that she experiences for the time being her own division: ("Am I myself or my brother?"). It is precisely in this connection that she subtracts him from the real number of her siblings. Herein lies the true drama of twins: it seems better to die than to be separated from the other twin. This lack has the significance of a mise en abyme for the little girl, configured in an inverted form (cf. the absence of the brother in the sibling group) and achieved by the difference between the numbers four and five.

Here is another example. "I was five and a half months pregnant. I didn't know who the father was, but I knew that I was separating from my husband. I was soon going to benefit from maternal leave, but I had forgotten to announce my pregnancy. I then went to work, only to be sent away again; I had been made redundant." The five and a half months of pregnancy are to be understood here literally, that is to say, as two times five. The associations confirmed that it referred to the birthday of her daughter, who was going to be 5 years of age on 5 May (which, moreover, is the fifth month of the year). Faced with the torments of the divorce, causing depression and morbid ideas, she returns to the time of her pregnancy. It was a period rich in events of high psychic value. While she was pregnant, she learnt that her father and her stepmother had decided to adopt a child! According to her, this confirmed not only the hatred that her stepmother felt towards her, but also the ambiguous relationship between her stepmother and her father.

I am now going to take one of Freud's dreams concerning numbers. "I received," he writes,

> a communication from the Town Council of my birthplace concerning the fees due for someone's maintenance in the hospital in the year 1851, which

had been necessitated by an attack he had had in my house. I was amused by this since, in the first place, I was not yet alive in 1851 and, in the second place, my father, to whom it might have related, was already dead. I went to him in the next room, where he was lying on his bed, and told him about it. To my surprise he recollected that in 1851 he had once got drunk and had had to be locked up or detained. It was at a time at which he had been working for the firm of T—. 'So you used to drink as well?' I asked; 'did you get married soon after that?' I calculated that, of course, I was born in 1856, which seemed to be the year which immediately followed the year in question.

(Freud, 1900, pp. 435–436)

The dream is characterized by this illogical temporal immediateness between the years 1851 and 1856. This succession is there, it may be supposed, to indicate the *symbolic* relation of cause and effect between the father's "death" and the son's birth. We saw earlier that paternity is combined with the symbolic death of the father.

I am not going to return here to Freud's interpretation of it. However, there is one element that is worth noting. Freud often makes comments on this dream when he is discussing others. Could this be the sign of an oversight, of something incompleted or not adventized? We can find some indications of this in the explanation suggested by Ernest Jones.

Referring to Freud's health problems in 1894 (see the letter from Freud to Fliess dated 19 April 1894 in Freud (1985)) and to an unpublished part of his correspondence with Fliess, Jones remarks on the superstitious number that appears in the dream (Jones, 1957, p. 341). According to the law of the permutation of numbers, a complete fabrication on Fliess's part (see Freud, 1900, pp. 166–167, footnote 2), if the numbers 23 and 28 were added up, the result, that is, the number 51, would be a bad omen. Now, Fliess allegedly predicted that Freud would die precisely at this age! Were these superstitious calculations really the work of Fliess or operations stemming from Freud himself, that is, from a man struggling with problems of health and who, furthermore, was under the influence of a massive transference of love towards an ingenuous friend? Can one rely on Jones' assertions? Is the connection he makes between Freud's dream and Fliess' predestined numbers not a pure invention? The answer, at least in part, does not seem difficult to find. It is in the interpretation that immediately follows the dream narrative that Freud himself mentions the elements that Jones provides us with. New indications of this can be found in other places in *The Interpretation of Dreams.*

In the section on the forgetting of dreams, Freud, taking up again the content of the dream in question, writes this:

In the apparently absurd dream which treated the difference between 51 and 56 as a negligible quantity, the number 51 was mentioned several times. [See 1900, pp. 166–167, footnote 2; pp. 449–450; p. 382 and p. 513.] Instead of regarding this as a matter of course or as something indifferent, we inferred from it that there was a *second* line of thought in the latent content of the dream leading to the number 51; and along this track we arrived at my fears of 51 years being the limit of my life.

(p. 513)

Numbers, as figures of finiteness, seem to approximate to what Lacan calls pure signifiers. In any case, it is by considering them as such that the analyst will succeed in obtaining the necessary associations concerning them, that is, by disregarding their signifieds and the mode in which they manifest themselves. By way of example, in the dream about the old man, we would have learnt nothing from the number eleven if we had confined ourselves to its immediate signified which presents itself in the form of hours. The relation between numbers and death[26] seems to be inherent to the oneiric drive which attempts to adventize the torments and worries of man. These pure signifiers are also bound up with what determines our finiteness.

Transformation of a fact into its contrary

Under this heading, we are going to study dreams that appear to run counter to the notion of adventization, even though the latter seems to take place from the point of view of analytic treatment. A young man of 32 years of age began an analysis in the hope of understanding the reasons for his professional failures. The analysis brought to light an unaccomplished task of mourning concerning the death of his father of whom he was the only son. His parents' divorce was one of the important events of his childhood. The care of the children, him and his two elder sisters, was entrusted to the mother who had the unfortunate habit of turning them against their father. Identifying with their mother's desire, the children continued, throughout their childhood, to have a negative attitude towards their father which turned into open conflict later on. It was following one of these conflictual scenes that the young man left his father, whom he never saw again. In this way he missed the last opportunity to say farewell to a father who, all the same, had always fulfilled his role.

"I dreamt that I was with my sisters in the mountains. We were at a table in a restaurant. Some hunters came in who had lost a member of their group. My sisters began looking at each other with complicity. I understood that they had hidden from me the fact that my father was still alive."

The dream is full of paternal signifiers. The young man's father ran a restaurant when he was alive and used to take his children to the mountains each year. The hunters with their berets evoked the members of the Resistance to which the father had belonged during the Occupation. He would tell the story of how he had once got lost in the snow, giving his fellow combatants cause for concern. This story of how he had got lost pointed to another more tragic one in the form of his tragic demise for the son.

The looks of complicity between the sisters were nothing other than a denial of the father's death, that is to say, as the analysis showed, the beginning of its integration within the mind of the young man. *Denial* is not *rejection,* but affirmation in the form of negation (see Movallali, 1988). The subjective place of the young man is located in the division involved in the affirmation of the father's experience of getting lost, driven away by the young man on the occasion of their last meeting, and the negation of his death. The missing object is not, however, the father whose death has now found its place in the son's negating confession, but rather in the knowing gaze that he might have cast on him when he was alive, and which he will never have the opportunity of doing again.

Here is another example of this kind of dream. A woman who was approaching the age of 50, and whose husband had left her three months earlier, had the following dream. "I was lying in a single bed when my husband joined me and wanted to lie down next to me. I was overjoyed. Then, we were together the whole time; we were working in the same place. At a certain moment, we were at the threshold of a doorway (of the place, no doubt, where we were working). Apparently for fun, my husband prevented a young woman from passing through the doorway. She asked him if they could go out for a walk one day together. He agreed. Then he had to leave me to go back home; perhaps he was living with someone. At any rate, he was living his life and that seemed normal. All these comings and goings in his company took place in a town located on a plateau. What was strange was that, in spite of feeling happy, I had a strong sense of regret, and even of sorrow, a feeling that has never left me since."

The town on the plateau reminded her of the expression "bringing everything on a plate (*plateau*)". Her husband had left her permanently. His return in the dream was "too good to be true". What we have here is a mild state of melancholy which perhaps announces the end of her suffering, an attempt to adventize what is finished for ever. Is the feeling of happiness an admission of the underlying sorrow? Or is it connected paradoxically with the relief that the process of mourning, already underway, promises her thanks to the adventization?

There is another type of adventization in the form of denial which frequently occurs in the course of an analysis. It concerns, and we shall see why, one of the crucial moments in analysis. I am referring to dreams that dramatize the castration of the subject through diverse forms of negation.

A woman saw herself in a dream endowed with masculine sexual organs; she was walking completely naked in the street. As her analysis showed, she was going through a tough period in connection with castration. Ever since the moment she saw, at the age of 15, her mother naked, she has constantly been somatizing. Her somatic disorders concern in particular the bodily region which she had seen. The dream in question does not have the significance of denial, but of adventization. To see it as a rejection of castration would be to lose sight of the main issue at stake in her analysis. Evidence of this is provided by the two following dreams which occurred a few sessions later. "I was in the Bois de Vincennes, my children were very small. My elder son was no longer there. The park was empty. I was screaming in vain." She then gave the account of the "*verbe*" (sic) as follows. Questioned about the slip of the tongue, she associated to the words of the Gospel: "In the beginning was the Word" (*Verbe*), and then added, "it's nothing but verbiage". Faced with her castration, symbolized in the dream by the distressing loss of her son, she is halfway between the sacred and the ridiculous, which is how she characterized the analytic situation. Now for the following dream: "A young girl has died; it's my eldest son."[27] The phallic place occupied by the son brings with it, in its adventization, the grieving sorrow of the young girl that she was for her mother. Her analysis shows that her mother's existence was a token of love for her in order to countervail her own "lack" as a woman. In her mother's love, she was constantly searching for the same excessive affection that the latter showed towards her brother.

Here are some other dreams from the same period. "I was in the street in my pyjamas, feeling ill-at-ease; there were women's clothes "parked" on the ground. My eldest son urged me to put my clothes on. I couldn't find my car." The phallic attribute (the car) takes the form of feminine objects capable of being removed and detached from the body. The presence of her eldest son in the three dreams attests to the phallic attribution that she enjoys by procuration. "A suspended bridge. On the other side, it's Germany – the Black Forest is in the distance. I didn't go there, I was afraid of getting lost." She added that the terrifying aspect of the dream was related to the huge void between the "two legs" of the bridge, and that Germany was indirectly her mother's land of origin. The maternal phallus, insofar as it is missing, is thus equated with the terror that the black continent inspires in her. And so it is rejected like the cake of the same name (*Forêt Noire*, Black Forest) that she detests. This orality showed all the ambivalence that tied her to her mother. It is clear that the overdetermination of the signifiers relating to loss simply confirms the adventization in process. The eclipsing of the father from the family scene had made the mother/eldest son pair the hard and uncastratable core that served our dreamer, by procuration, as "more than enjoying" (*plus-que-jouir*; Lacan), the lack of which proved to be a hard ordeal.

Following her menopause, a woman had the following dream. "I had three children (but only two in reality). On the ground floor there was a fire; I went down and put the fire out with a fire-extinguisher." It is possible that the dream was triggered by a hot flush, a postmenopausal phenomenon from which she was suffering at that time. Dreams are above all the guardians of sleep. The sensation of heat gives rise to the use of the fire-extinguisher. Certainly, the fire is the dream's depiction (*Darstellung*) of the disagreable sensation of heat in order to prevent the dreamer from waking up. Nevertheless, as soon as it appears in the dream, it assumes an adventizing configuration. The dream concerns, as we shall see, castration in the form of negation, the fire-extinguisher standing in for the missing phallus.

"I had three children instead of two", was a sign of the negating adventization of this castration that the recent hormonal changes represented for her, undermining her sense of femininity. She had the tendency to use this as a means of exerting power over men.

The question of the menopause simply gave more force to the adventization of the lack of the phallus taking place in her analysis. Her history showed that she had a strong penchant for seducing men, seduction, in her case, that stood in for phallic demand. She described herself as a "tomboy" who knew how to handle men.

Seduction can go hand in hand, in a woman, with what may be called the phenomenon of the veil (see Lacan, 1960–61). The latter has the function, then, of showing, while seeming to hide what is involved. The woman thereby creates the illusion that she has it. By dissimulating, the veil makes one believe that she really possesses the phallic attribute. There are even societies where it is men who take charge of this in an effort to address their own fear of aphanisis (disappearance of one's sexual desire). The phenomenon of the veil is obviously not unrelated to feminine narcissism, which as Freud pointed out, exerts extraordinary power over men. One of the forms of behaviour linked to the phenomenon in question is that of the *femme fatale*.

The *femme fatale* is not only a mode of phallic demand, but also a curious ambiguity towards the man. So what determines her mode of being depends more on her intrinsic relation to the veil than on her supposed beauty.

Let me cite another dream of our analysand. "I had my genitals outside of my clothes and everyone could see them. I was extremely embarrassed and tried, in vain, to hide them." As a result of these dream events, she decided to put an end to her equivocal attitude towards one of her colleagues whom she had been keeping in suspense up until then. During one of these scenes of seduction, she had accepted to dance with him in a physically close and provocative manner; but, a moment later, she was pushing him away, quite aggressively, reproaching him for getting ideas into his head.

The ultimate invocation of desire

There are dreams where desires that have already been adventized make their last appearance. In this ultimate manifestation, we fear that in spite of clinical gains, the conflict in question has resumed with greater intensity. But the strength and acuity with which the adventized desire reappears on the stage barely hides its negating character.

Here is an example of this type of dream. The dreamer had to endure a lot of suffering after her husband had left her. In analysis she went through an intense work of mourning followed by a long period of sorrow. "I was six months pregnant. I calculated the time that was left until 11 November, the day on which I was supposed to give birth. Things happened more quickly than expected and I was going to give birth well before the estimated date. I expelled the baby, but was convinced it was dead. The baby started screaming and I called to my (ex-) husband: 'the baby's alive!'" The dream needs no commentary. However, the numbers call for clarification. The 11 November is not only the date of the armistice between the belligerents ("between her and her husband"), but is also the month of the dreamer's birthday. The number six alludes to the schedule of the divorce. The meaning of the dream approximates to the case of another woman who, when she left her companion, reproduced all the somatic symptoms that she had during the first months of her pregnancy.

Below is a dream of a young woman, smitten with homosexual love. Like her two brothers, she was wounded by a divorce that had shaken all three of them up; it was a real trauma. The terrifying dissension of their parents had, until their father's death, destabilized the entire family and shattered the ties between the children.

"I dreamt about my mother. She had a Jewish name, Céline Rosarhom. She was extremely beautiful and the name suited her wonderfully." Some associations followed. Céline made her think of the eponymous song that begins as follows: "Why have you never thought of getting married?" The second component of *Rosarhom* made her think of the word *man (homme,* in French). Then, she wondered if Céline sounded Jewish. In view of the anti-Semitic work of Louis Ferdinand Céline, this last remark took on particular significance. The allusion was immediately subject to the effect of censorship leading to a confusion regarding the subject of the writer's origins. In the session preceding this dream, she had mentioned her maternal grandmother who had experienced the humiliations suffered by collaborators after the war. "*Why have you never thought of getting married?*" was a question addressed to her mother, which converged with her omnipotence (you don't need to get married, you don't need a man). In the mother's omnipotence, there was at once feminine essence (Rosa) and the signifier *man.* The same elements can be found in the pseudonym of the writer, Céline, a name he borrowed from his mother. The mention of the pseudonym of the anti-Semitic writer in the dream underscored the extreme ambiguity within the family. A Jewish father, who was in the French Resistance, and who had married the daughter of a mother accused of collaborating with the Germans; a woman who opposed her mother, who was a collaborator, and married a Jew, but who did everything possible to denigrate him as the father of her children . . . these were all ambiguous elements that must have destabilized the family as a whole.

The analysis established that the dream was an attempt to adventize the loss of the passion that she had had since her early childhood for her omnipotent mother. It was also an attempt to rehabilitate her father, a Jew and a member of the resistance. It was not Céline that sounded Jewish, but *Rosarhom.* The charming aspect of her mother in the dream merely betrayed her lack, thereby undermining, like a last flame of desire, the omnipotence that was so revered by her daughter. The mother seemed to serve her, in the dream, as a projective mise en abyme. She had to represent her own person in a narcissistic form ("She was extremely beautiful and the name suited her wonderfully"). The mother's "Jewish name" stemmed from an old phantasy of seeing her parents get together again. Rosarhom combined *Rose,* a narcissistic attribute of the mother, and *hom* referred to her father. The letter R in her name (Carole) joined the parental signifiers. It goes without saying that the dream spoke throughout, albeit in an inverted narcissistic form, about the mother's lack, the object of the adventization. She thus took leave for the last time of her mother's omnipotence.

The following example is also related to separation. It concerns a young man who had a love affair with one of his work colleagues. "I was working at . . . X, but elsewhere at the same time. I had forgotten a cassette on my ex-girlfriend's desk, or else she had walked off with it. I was going to her office to recover it. She had left me a message on a newspaper (as she often did): 'Nothing is better than holidays,' signed *mie de pain.*" They had in fact begun their love affair while on holiday. *Mie* evoked for him the word of love of troubadours and *pain* was related to the major signifier of the ex-girlfriend in question. As for the cassette that had disappeared, it expressed the dreamer's unhappy love affair with his ex-girfriend. They had in fact avoided divulging their love affair to their colleagues. The message left by his girfriend on the newspaper was the last sign of tenderness that the dreamer demanded from her while announcing, with regret, the end of their adventure.

In the last example, below, we are not dealing, as in the previous dreams, with the last flame, like that of a desire, but with the *mourning* that accompanies it. The dreamer is a teacher at the end of his career returning from an important job abroad. Back in his own country, he found himself practically out of work. A period of sorrow followed and a loss of self-confidence. The dream, which needs no commentary, occurred at a moment when he finally received an offer of a job completely at odds with his competence and his aspirations. "I was in a big hall, like in a university, in the presence of a large audience. I didn't recognize any faces except the members of my family whom I hadn't seen for a long time; in the first row, there was my cousin Didier. I felt the words becoming heavier in my mouth. I now began to stutter, but continued my speech all the same. I was getting more and more tongue-tied. I started shouting in front of an imperturbable public, who, moreover, were inert and indifferent. I woke up sweating profusely."

Notes

1 As for Damasio, he finds the self (*moi*) in the different levels of cerebral functioning. In his recent book, *The Feeling of What Happens: Body and Emotion in the Making of Consciousness* (Damasio, 2000), he distinguishes three registers of the ego: the proto-self (our phylogenetic heritage), the core- (or current-) self, and the autobiographical-self, each of which is governed by homeostasis. According to him, homeostasis is a perpetual act of struggle to reach a state of equilibrium. But this struggle is conceived of in terms of the concept of *survival* and not in terms of the internal conflicts of the human being. As can be seen from his recent book, *Looking for Spinoza: Joy, Sorrow and the Feeling Brain* (2003), Antonio Damasio believes that the mind strives, in conjunction with the body, to preserve itself in order to embrace happiness. The truism of the neurologist seems to disregard the fact that the real struggle in us as mortals combines both destructive forces and vital power, which are both engaged in a constant conflict whose supposed outcome, namely happiness, is greatly compromised.
2 The recent debate on the new human subject – insofar as he is said to have fallen into the narcissistic rather than in the symbolic and is experiencing an unprecedented crisis of identity due to the decline of the father abolishing his own place of subjection to the Other – should take this distinction between subject and subjectivity into account. In this connection see in particular Charles Melman and Jean-Pierre Lebrun (2009).
3 See also the Arab *arkh*, which is the root of the word *tarîkh* (history).
4 It is called paradoxical owing to the great influx of cerebral activity that is triggered during the stage of sleep in which dreams generally occur.
5 A system set up by Jacques Lacan with the purpose of defining the end of analysis, settling the question of the didactic analysis, and proceeding to the nomination of psychoanalysts.
6 Translator's note: a play on "*en avant*": ahead of, in front of.
7 Μίμεσις is the term with which Plato defines a narrative related in the first person. It does not belong, according to him, to the art of poetry where the poet relates events and

reports indirectly what the characters say. Aristotle reverses this relation and situates poetry under the same banner as other artistic activities. Arab authors of the Middle Ages had combined these two conceptions while reserving the word *hekâya* (literally *μίμεσις*) for their fables as a whole. "Please refer," Henry Corbin (1971) writes, "to what is said here concerning the Arabo-Persian term *hikâyat*, which can be the source of inexhaustible meditations because it has the virtue of connoting at once the idea of history and the idea of imitation (the Greek mimesis)" (p. 18). It goes without saying that here imitation certainly does not have an imaginary sense. In the sense in which Arab authors understand it, *hekâyat* is the symbolization (the imitation) of history as such. "In the act or the action of the narrative, of the "recital" (the *hekâyat*), the story-teller, the action recited, and the hero to whom the narrative relates or the person who tells the story, are one and the same person" (ibid., vol 2, pp. 202–203). Hence the extraordinary entanglement of magical enchantment and reality that we find when read-ing this literature. Henceforth we will recognize it as containing what seems to be at work in dream activity.

8 *Abyme* (masculine noun, from the Greek *abussos*, bottomless). *En abyme*, used of a work that is cited and contained within another work of the same nature (narrative within a nar-rative, painting within painting etc.). (Bibliorom Larousse, version 1.0, 1996.)

9 Cf. Freud: "Dreams are the guardian of sleep" (1900, p. 233).

10 Topography refers to the distribution of the psychic impulses in different agencies of the mind. In his first topography (*The Interpretation of Dreams*, 1900), Freud distin-guishes the three systems which, according to him, constitute the psychical apparatus, i.e. the conscious, the preconscious, and the unconscious. In his second topography (*The Ego and the Id*, 1923) he divides them up into three places: the id (*das Es*) the ego (*das Ich*), and the superego (*das Über-Ich*).

11 For a full discussion of the use of the terms "phantasy" or "fantasy", see the article in Laplanche and Pontalis, 1967, pp. 314–318.

12 Secondary revision (or elaboration): "Rearrangement of a dream so as to present it in the form of a relatively coherent and consistent scenario" (Laplanche and Pontalis, 1967, p. 412).

13 Screen memory: "Childhood memory is characterized both by its unusual sharpness and by the apparent insignificance of its content. The analysis of such memories leads back to indelible childhood experiences and to unconscious phantasies. Like the symp-tom, the screen-memory is a formation produced by a compromise between repressed elements and defence." (Laplanche and Pontalis, 1967, p. 410–411)

14 Deferred action, *Nachträglichkeit*, Fr. *après-coup*): "Term frequently used by Freud in connection with his view of psychic temporality and causality: experiences, impres-sions, and memory traces may be revised at a later date to fit in with fresh experiences or with the attainment of a new stage of development. They may in that event not only be endowed with a new meaning but also with psychical effectiveness." (Laplanche and Pontalis, 1967, p. 111)

15 Translator's note: "*à pierre*", suggests both that the pistol fires stones and that it belongs to Pierre.

16 Translator's note: In French, an I.M.E (*Institut Medico-Educatif*).

17 Depersonalization is an almost current phenomenon in adolescents in search of their identity. It consists in a slight alteration of the image that the young person has of him or herself, and may even involve a certain sense of transitory de-realization. The young person contemplates his or her image in the mirror and sometimes ends up harm-ing it by attacking it. Alternatively, he or she may ruminate over the various dangers that could harm his or her physical integrity. The depersonalization in question here is fundamentally different, both in intensity and in terms of mental affliction, from the discordance that can be observed in psychotic affections.

18 All nightmares are horror dreams, but not all horror dreams are necessarily nightmares. What distinguishes the latter is that they wake the subject up and interrupt the course of his dream formation. They arrive at a point where the subject is no longer able to

integrate them into his world. The threshold in question is the moment when the subjective world meets that which *ex-sists* it as real. The latter is, so to speak, the "contour" where our symbolic world is constituted. This frontier inevitably remains inaccessible to us. The slightest sign of its manifestation is a source of terror.

19 Françoise Dolto (1984, p. 153) speaks of "deadly fascination".

20 Iterative: "Relating once what has happened once (nR/1H) . . . this type of narrative, in which a single narrative emission comprises several occurrences of the same event, is what I call an iterative narrative" (Genette, 1972, pp. 147–148).

21 Singulative: "Relating once what happened once (or, if one wants to abbreviate this into a pseudo-mathematical formula: 1R/1H) . . . This form of narrative, where the singularity of the narrative statement corresponds to the singularity of the event narrated, is obviously by far the most common, and apparently considered so 'normal' that it does not have a name, at least in our language. . . . I propose to give it one: henceforth I will call it singulative . . ." (Genette, 1972, p. 146).

22 The primary process is a mode of functioning of the psychic apparatus identified by Freud in which "psychical energy flows freely, passing unhindered, by means of the mechanisms of condensation and displacement, from one idea to another . . ." (Laplanche and Pontalis, 1967, p. 339).

23 *Jouissance*, for Lacan, is the very lack in the Other to which the subject is called on to remedy. It is beyond the opposition pleasure/unpleasure and is marked by its repetition and persistence.

24 See further on, the third chapter: "Determinism or pure chance?"

25 Translator's note: In French *ivre* means drunk.

26 Or other signifiers that are related to it, birth, baptism, marriage, paternity, giving-birth, adoption, and even moving house and the purchase of property . . . all of which constitute major acts during our lifetime.

27 This was followed by the association of an unidentifiable fairy-tale: "A father loses or abandons his daughter in a wood; she is recovered by a crow . . .".

7

DREAMS AND THE NEUROPHYSIOLOGY OF SLEEP

The sleeping brain is not part of a reflex system

I am going to turn my attention now to sleep, which supposedly reveals other characteristics of dreams, in particular, their strangeness and incongruity. We will see if these are due to the specific state of sleep at the neurological level or if they depend on the dream work itself. Dreaming as mental activity depends on sleep. Is there a close articulation between sleep, a neurological process, and dreaming, that high place of human subjectivity? Do the somatic and the psychic stem from a dichotomy, from an irremediable divorce? How, in the light of recent discoveries in neuroscience over the last century, can one apprehend Freud's contribution to the understanding of man? In this connection, we will have to turn our attention to the neurology of sleep in order to examine the place and functioning of oneiric activity.

The refutation of the brain as a reflex system was, in fact, the first decisive step in the new research in neurophysiology. These investigations were able to prove, in the most convincing manner, that the brain does not cease its activity in the absence of stimuli from the external world. On the contrary, it stimulates itself spontaneously, that is to say, endogenously.

The correlation that Aserinsky and Kleitman (1953) discovered between rapid eye movements (REM) and paradoxical sleep served in particular to confirm this new approach. "While scientists have yet to compare," Alan Hobson (1988) writes,

> the response of a single cell to a visual input from external world during the wake state with the activity of that same cell in REM sleep when the visual system is internally stimulated, activation-synthesis predicts that the patterns of response to both internally generated and externally generated stimuli will be remarkably similar. A scientist who has only the cell's

electrical response pattern to judge may even find it impossible to predict whether an animal is processing external or internal visual information.

(p. 165)

The entire Pavlovian theory of the nervous system was, as we saw in the first part, based on the converse thesis. It coincided with the popular idea that the brain regularly interrupts all forms of stimulation in order to rest and renew its resources of energy. This concept of *de-afferentation* was given a more elegant explanation by the English physiologist Charles Sherrington. For him, it was not the fatigue due to the repetition of conditioned responses that led the brain towards rest. He believed, on the contrary, that cerebral activity diminished greatly during sleep before ceasing gradually. For Sherrington the brain was like a *magic loom* whose shuttle movements illustrated its stimulations. These were supposed to transform themselves into small lights at the onset of sleep, tending towards quasi-complete extinction for the rest of the night (Sherrington, 1942). But we know that this is not the case. Man is never cut off from the world that inhabits him so intimately. By intimate, I mean the highest degree of dissension and discord. This conflictual state of affairs is the principal fabric of the oneiric world on which desire and adventization are perpetually being woven. The brain is not switched off during sleep. Its immense network of activity is at its height during paradoxical sleep, and can even surpass the intensity that can be observed in it in the waking state. Freud had already demonstrated this in his *Interpretation of Dreams* in 1900. The dream constitutes the privileged site of disputes and conflicts, underpinned by the cerebral activity that is sleep.

The first recording of the individual cells of the brain was made by Jasper and his colleagues in 1957 (Jasper et al., 1958). It is true that it contradicted the idea that brain activity was marked by a lowering of intensity during sleep but, at the same time, it brought other unexpected results. Notwithstanding the slow waves recorded during sleep at the level of the electroencephalogram (EEG), investigators noted the presence of brain activity in many cortical neurons that was just as intense as during the waking state with their characteristics of weak amplitude and fast waves, reserved for active neurons. As for the neurons that are inactive on waking, they remained so in spite of intense and sustained waking activity, but on the contrary became active again during sleep.

Hubel and Evarts (see Akert et al., 1965) carried out more systematic studies concerning the recording of individual neurons. They applied them primarily in the domain of the neurophysiology of sleep. Their investigations put a definitive end to reflex physiology, which defended the thesis that the brain is at rest during sleep. The researchers were particularly interested in a much more active stage of sleep which seemed to be the most conducive phase for dreaming proper. Hence the term ascribed to it by Michel Jouvet, the famous French neurobiologist, of *paradoxical sleep*, the stage supposed to be the privileged locus of oneiric activity. As for Dement (1958), the American neurophysiologist, he had referred to it as *activated sleep* as early as 1958.

As soon as we close our eyes

There is a singular phenomenon that occurs as soon as we leave behind our waking activities. I shall baptise it henceforth as *oneiric automatism*. Having our eyes closed gives rise to a set of imaginary activities ranging from simple transitory images to the formation of oneiric activities, including other manifestations as well that remain to be identified. The imaginary unity that I have referred to as a *flash* can also manifest itself before the definitive phase of slow sleep. The reveries that occur as one is going to sleep and the images or sequences formed afterwards are all oneiric elements that are likely to be the object of investigations in order to distinguish their characteristic traits.

Oneiric automatism is the confirmation of what may be described as a primordial entente between man and the world. This primordial entente links us so closely to the external world that we continue to have access to it even in its absence. It is the "interior" gaze of man always-already open to the world. It is the condition of this openness. Neuroanatomy confirms this. The retina, situated at the back of the eye, registers the visual stimuli passing through the crystalline lens. They are inverted by the latter. Having reached the back of the eye, they are transformed into electric impulses. The optic nerve then transmits them to the optic chiasma, where they cross. The part of the thalamus called the lateral geniculate nucleus processes them in turn before sending them to the primary visual cortex in the occipital lobe.

In the absence of external visual stimuli, that is, when the eyelids are closed, the trajectory described above is inverted, giving rise to eye movements which reach their highest level of intensity during the stage of *paradoxical sleep.* This inversion might explain the genesis of dream images that occur in the absence of external stimulation. The explanation would have no basis without recourse to the concept of a primordial entente between man and the external world to which I have just referred. As Michel Jouvet (1993) writes:

> Study of latency between the arrival of signals in the visual cortex and eye movements revealed a paradox. In the alert, awake animal the retinal signals provoked by a target object arrive in the visual centers before pursuit eye movements begin (the cause precedes the effect). In contrast, in the dreaming animal the beginning of eye movements *coincides with* or *precedes* the arrival of the endogenous nonretinal signal (the PGO[1] activity) in the visual cortex. It is obviously impossible that the effect (exploratory eye movements) should precede the cause (visual hallucination).
>
> (p. 89)

Such a phenomenon finds an explanation in the theory of emergent properties, founded on the principle of a prior entente between man and the outside world. This theory claims to transcend the dichotomy between objectivism and subjectivism. It pleads for what Francesco Varela calls the middle path. "Cognition," he writes, "far from being the representation of a pre-given world, is the combined enactment of a world and a mind starting from the history of the diverse actions accomplished by a human being in the world (Varela et al., 1993, p. 35).

The oneiric automatism could be considered as belonging to these emergent properties insofar as it is akin to the self-organizing model of which Francisco Varela speaks. It emerges, in fact, from the "coupling" of the day residues with the historical coordinates that are the memories aroused by them. Like the connexionist model in question, it results each time, Varela (1996b) writes,

> from the birth of a "world" for the system that it causes to emerge from aleatory circumstances, in the course of the history of the coupling. There is, of course no "representation" of this sub-set of sequences selected among all the possible aleatory sequences of the system, and nor was it its task to recognize it. It is simply the existence of the system itself that has caused them to emerge from an indefinite mass of possibilities.
>
> (p. 106, translated for this volume)

An emergent property is what is constituted spontaneously during an encounter between two entities that share a certain number of elements belonging to their history. Its emergence occurs in the same way as "a non-existent path that appears while one is walking" (ibid, p. 111). We will see further on[2] how the aleatory dimension of emergence has its place in the overdetermination of oneiric activity.

Oneiric logic and paradoxical sleep

We have seen the major role played by the interacting bridge established between the thalamus and the cortex in visual stimulation (exogenous and endogenous). It is with its activity of automatic spindle detection that the same cortico-thalamic part seems to deprive us of perception and consciousness during sleep (see Llinás and Paré, 1991). This is the beginning of the first stage called slow-wave sleep, during which cortical activity gradually slows down. This slowing down also concerns the consumption of glucose by the visual cortex, brain temperature and postural tonus. Eye movements may exist at this stage, but their presence does not seem to be constant in any notable way. Given the correlation that seems to exist in sleep between eye movements and the presence of dream images, the relative insignificance of these movements during slow-wave sleep (deep sleep) will easily be understood. There is reason to think that, during this stage, the occurrence of dreams is less marked than during paradoxical sleep, the privileged locus of oneiric activities. We may also suppose that dreams devoid of oneiric logic tend to occur during this first slow-wave stage.

The high complexity of the dream, its narrative structure as a dramatic sequence, its condensed texture, its turbulent intrigue, its recourse to linguistic devices that confer on it its overdetermined character, its attempt at adventization, in short, the components of the oneiric logic are all elements that run counter to a tranquil mind, far removed from serious preoccupations and at rest.

It is thanks to the research that has been undertaken since the mid-1950s that we know that there is a dream stage during which brain activity is at its highest level of intensity, contrasting radically with any period of rest or relaxation. Hence the term *paradoxical* sleep. This intense brain activity also affects, particular in the case of strong emotion, the cardiac or respiratory frequency. As we can easily imagine, this could be related to the dream intrigue. Not only does the consumption of glucose by the visual cortex – which diminishes initially during deep sleep – increase, but so does that of the cortical activity situated at the level of the cortex and eye movements. "It thus seems," Jouvet (1993) writes, "that dream consciousness uses more energy than waking consciousness" (p. 52).

The discovery of paradoxical sleep confirmed, once again, the pertinence of the Freudian discovery. Only such a neurological flurry of activity could provide backing for the overdeterminism and remarkable complexity that Freud ascribed to the dream as a royal road to the unconscious.

Neurophysiolgical mechanisms of paradoxical sleep

Neurobiologists have been able to identify the cerebral system that is responsible for paradoxical sleep (see Dement, 1965). Jouvet qualifies it as endogenous and specifies it as PGO activity. This is a neurophysiological system that concerns almost the whole of the encephalon. PGO activity (ponto-geniculo-occipital waves) concerns, as its name indicates, both the visual centres (occipital cortex) and what serves as a relay for visual and auditive paths (geniculate bodies). The reticular formation that extends from the pons, situated in the brainstem, to the thalamus provides the PGO system with a tremendous extension of its activities at the level of the cortex (see Figure 7.1).

Owing to their role as audio-visual relays, the geniculated bodies, participating actively in PGO activities, have great importance for psychoanalytic theory. We know, since Freud, that representability (*Darstellbarkeit, figurabilité*) is one

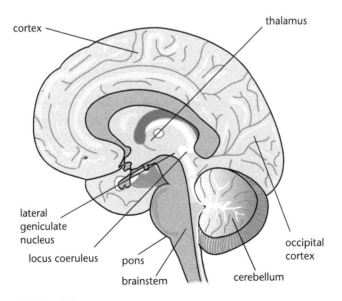

FIGURE 7.1 PGO activity

of the principal mechanisms of oneiric activity. It is by virtue of this device that words and linguistic expressions are transformed into images and that the dream requires interpretation. In the experiments conducted on paradoxical sleep, the emphasis is placed essentially on the importance played by visual phenomena, whereas more recent investigations also seem to take account of auditive phenomena. Velluti (2010) writes:

> [The] somatosensory and visual neurons exhibit changes in their firing rates in correlation with stages of sleep and the waking state. This is consistent with the hypothesis that a general shift in the neuronal network/cell assembly's organization is involved in sensory processing that occurs during sleep. This assumption is supported by magnetoencephalographic (MEG) study of auditory stimulation during sleep performed in humans. . . . Most important, no neurone belonging to any auditory pathway level or cortex was observed to stop firing during sleep.
>
> (p. 145)

This differs from the thesis of Michel Jouvet (1993) who, drawing on his own dreams, thinks that vision and audition are mutually excluded in oneiric activity. "Our EEG measurements," Pierre Etevenon writes, "have also lent support to [this] hypothesis . . . There would thus be a sort of preponderance of vision or of audition in the dream, one of these two sensory modalities dominating the other."[3]

We must not lose sight, however, of the the fact that it is sleep that constitutes the object of the neurophysiologist's research, and not the dream. But, however distinct from each other they may be, one must be careful not to get caught up in a dichotomy, thereby losing sight of the intimate relations between them. This is what I am trying to keep in mind here.

The existence of REM is one of the major characteristics of paradoxical sleep (PS). Researchers think that a direct link can be established between these movements and the occurrence of dream images. "The main external signs of paradoxal sleep," Michel Jouvet writes,

> are postural atonia and rapid eye movements, whereas the internal signs are cortical and PGO activity . . . Postural atonia can be triggered by exciting various systems. During paradoxal sleep we can record increased activity of neurons in many structures, but often this activity is related to associated phenomena such as eye movements.
>
> (1993, p. 45)

This is where, owing to a curious mechanism, a block in the motor neurological system occurs at the level of the spinal cord. This *postural atonia* is commanded, during paradoxical sleep, by a group of neurons in the part called the *locus coeruleus,* which is located at the extreme upper point of the brainstem, in the vicinity of the thalamus and the reticular formation (see Figure 7.1). By intervening at the level of the *locus coeruleus*, that is to say by unblocking it, one can activate it and provoke *oneiric behaviour* during paradoxical sleep. This is when the subject who is fast asleep may stand up and enact what he is dreaming. This experiment was conclusive in cats (Jouvet, 1993). Deprived of language, the cat's dreams only enacted what characterizes it in terms of instinct, that is, attack behaviour, being on the watch, pursuit of prey, etc.

The activity of the PGO system is most probably not limited to the excitation of the oculomotor muscles. It extends to a large part of the reticular formation. This extends from the spinal cord, at the beginning of the brainstem, to the thalamus, before projecting itself via its own paths and relays on to the deep part of the cortex.

The current state of research

Notwithstanding the immense amount of research carried out over a period of more than 60 years now, the neurophysiology of sleep still does not seem conclusive. The first controversy concerned the direct link that the first investigators established between paradoxical sleep (REM) and dream formation. They would wake subjects up when low-amplitude and high frequency EEG waves were triggered, to see if they coincided with dream formation. Other investigators set about doing the same without the releasing of EEG waves. They discovered that the dreams formed during other stages of sleep (non-REM) did not correspond entirely with the description of their colleagues who described them as being a poor and discursive in character (see Foulkes and Fleischer, 1975). Increasingly numerous publications running counter to the equation between paradoxical sleep and dreaming led a certain number of investigators to abandon this hypothesis, even though it was the prevailing hypothesis in the 1970s. The further we moved away from the discovery of paradoxical sleep in the 1950s, the more the number of NREM dreams, which was initially very low, began to increase substantially, reaching about 50% of the dreams reported by different authors. The supporters of the equation REM=dreaming then considered distinguishing dreaming and dream memory, putting forward the hypothesis that NREM dreams are only memories of those that occurred during the earlier episode of REM. Without taking their inquiries any further, some investigators tried to attribute NREM dreams to the fabulations of subjects after waking up in the laboratory (see Nielsen and Tore, 2003). This is no doubt one of the symptoms that we come across in the research into the physiology of the brain, a symptom that involves an obstinate refusal to take account of the specificity of the human subject. Just as Charcot's patients produced hysterical symptoms publicly in order to conform with the master's teaching, so we would be justified in wondering about the validity of the objectivity of the research projects in question. It is astonishing to see that investigators do not reflect more on the nature and impact of *their* demand on the subjects from whom they expect to receive oneiric contents.

Research orientations in the physiology of sleep

Model of brain asymmetry

The model of brain asymmetry seeks to attribute the specificity of dreams, their discrepancies and bizarre features, to the functional difference between the left and right sides of the brain. To the extent that the visual image is one of the principal characteristics of dreams, it would be legitimate, according to the adepts of the model in question, to suppose that the functioning of the right side of the brain prevails over the left side. The right side is thought to be more suited to accomplishing the functions relating to the image, while the left side is thought to be confined to the linguistic and intellectual register.

Broughton is the principal investigator to have promoted the thesis concerning the role played by the two hemispheres of the brain in the alteration of the state of consciousness during sleep. It is a theory that could be considered as the extension of Kleitman's hypothesis, put forward in 1963, that there exists a basic cycle, throughout the wakefulness/sleep period, which is repeated every 90 minutes (Basic Rest-Activity Cycle, BRAC). For Kleitman, the BRAC was based on a phylogenetically more primitive and older model than the digestive rhythm which was adjusted to the organism's nutritive needs (see Kleitman, 1963). Broughton (1975) attributes these rhythmic fluctuations to the functioning of the hemispheres. The right side of the brain, according to him, is responsible for REM, and the left side for NREM. Expanding on Kleitman's thesis on the BRAC, he argues that the alteration of the state of waking consciousness is also the reflection of the to-and-fro rhythm between dreaming and sleeping. The research published by Goldstein et al. (1972) on seven persons as well as their four cats and five rabbits confirmed that these changes recorded in the brain hemispheres were concomitant with the rhythm of dreaming and sleeping. On the other hand, no correlation was found concerning the period of wakefulness (Sugerman et al., 1973).

However, no subsequent investigations confirmed the thesis of oscillation of the brain hemispheres as being concomitant with the rhythm REM/NREM (Antrobus et al., 1978). Reproducing the researches of Goldstein et al., Antrobus et al., only obtained contradictory results (Hirschkovitz et al., 1979). Sometimes they observed the prevalence of both the left and right sides of the brain, and sometimes that of the right side during sleep (REM or NREM).

Model of emergent properties

Current research seems to have gone beyond the naive hypothesis that consists in ascribing a split and rigid mode to the functioning of the two sides of the brain. They are closely linked with each other not only by the intermediary of the *corpus callosum*, but also by other cerebral structures. This powerful dynamism renders obsolete any suggestion that contravenes their interrelationship, a principle challenging not only neuroanatomy but also brain biochemistry.

Although the idea of linking the oneiric configuration to brain asymmetry was simplistic and devoid of any proven scientific foundations, it was so attractive that a certain number of psychoanalysts attempted to see it as a confirmation of Freudian theory (see, in particular, Pommier, 2010). Not only did Freud distinguish word-presentations from thing-presentations, but he also established the principle of representability (*Darstellbarkeit, [Fr.] figurabilité*) according to which words are transformed, during dreams, into images. Can dreams really be explained, then, by the prevalence of the right side of the brain which, in view of the putative deficiency of the left side during sleep, takes over our thoughts and preoccupations? This is clearly not the case. The dream configuration continues to raise questions for us and to stimulate new research. Such research requires, here, as elsewhere, a rapprochement between psychoanalysis and the neurosciences. Without such a perspective, psychoanalysts will continue to accuse this research of being reductionist, just as the specialists in neuroscience will probably persist in their point of view that is external to psychoanalysis. It is true that neuroscientists often have a tendency to apply their scientific literalness to the Freudian discovery, a discovery that is still, notwithstanding the immense clinical progress that has been made, at a very early stage in the understanding of human nature. Psychoanalysts will receive credit for not applying the same literal approach to their own scientific domain.

The theory of emergent properties has already greatly contributed to this kind of rapprochement. In this theory, neurons are considered as belonging to large ensembles. The latter appear and disappear constantly, following the interactions of the neurons which change and undergo modification not only in accordance with the *context* but also with the *tasks* to be accomplished. According to the adherents of this theory, all mental activity mobilizes both local and global brain networks. This gives rise to the cooperativeness of the subsystems covering the whole of the central nervous system. And so *emergent properties* are formed which introduce patterns of experience in close affinity with the surrounding world.

> As a result the entire system acquires an internal coherence in intricate patterns, even if we cannot say exactly how this occurs. For example, if one artificially mobilizes the reticular system, an organism will change behaviorally from, say, being awake to being asleep. This change does not indicate, however, that the reticular system is the controller of wakefulness. That system is, rather, a form of architecture in the brain that *permits* certain internal coherences to arise. But when these coherences arise, they are not simply due to any particular system. The reticular system is necessary but not sufficient for certain coherent states, such as wakefulness and sleep. It is the animal that is asleep or awake and not the particular neurons.
>
> (Varela et al., 1991, p. 94)

It can be seen, then, that the theory of emergent properties turns its back on the fragmentary view that is in vogue in the prevailing trend of neuroscience. Such theories will no doubt ring the death knell of the cognitivist movement which so dominated research in neuroscience during the twentieth century. Just as behaviourism obstinately refused to open the "black box" of the encephalon in its research activities, likewise cognitivism is turning its back today on any attempt to question the subjectivity of living things. Having arisen from cognitivist movements, the theory of emergent properties has succeeded in opening itself up to other currents of thought (see Cuvelier, 1992), including psychoanalysis (see Cohen and Varela, 2000).

Thalamo-cortical model or temporal loop of oneiric activity

The gap between neurophysiological research and subjectivity seems to be closing according to an article published by R. Llinás and his colleague, D. Paré (1991). We are now familiar with the theoretical approach of Rodolfo Llinás, after examining it in the first part of this book. According to him, the brain is a closed system, but in conformity with the reality of the outer world. This runs counter to the thought of William James which is at the basis of the functionalism of neuroscience today. The brain is not a system through which sensory input can generate "the functional scaffolding required to create an internal image consistent with external reality" (Llinás and Paré, 1991, p. 522). For Llinás, the brain is equipped, from birth, with an ensemble of innate structures that function in the same way as a priori categories at the philosophical level.

If the encephalon is a system that is essentially closed, paradoxical sleep must be considered as its most natural state. Indeed, it is during the latter that the brain is cut off from the outer world and finds itself in the greatest conformity with the its configuration as a closed system. Research undertaken since the middle of the twentieth century suggests that the only difference between the state of wakefulness and paradoxical sleep lies in the presence or absence of sensory afferents. In other words, there is fundamentally no difference between the two states. Being equipped with its own intrinsic functioning, the brain *modulates itself,* in the waking state, according to the sensory parameters of the outer world. Diverse studies show that only a minor part of the connections of the thalamo-cortical system is concerned by sensory information, even though the system in question is responsible for states of consciousness, in waking life as in the world of dreams (ibid., p. 522). The importance and autonomy that these authors accord to paradoxical sleep puts them in step with Michel Jouvet who ascribes a third state of consciousness to paradoxical sleep in addition to those of wakefulness and slow sleep. On the other hand, they do not share Jouvet's view on muscular atonia. "Does the lack of behavioral response to supra-threshold sensory stimuli reflect the somatic paralysis characterizing REM sleep, or rather, a difference in the way the brain processes sensory input? The latter explanation seems to be more likely" (ibid., p. 524). It is true that surgical intervention at the level of the pontine part of the brainstem prevents muscular atonia, revealing for the observer the dream content of the sleeping animal, but the latter continues to be insensitive to any outside stimulus. In fact, it is in a state of *brain alteration* and not of muscular paralysis as Jouvet thought. Hence the conclusion of these authors that paradoxical sleep results from a *functional change* at the thalamo-cortical level. However, this is a change in the state of the brain that remains faithful to the sensory data. This reminds us, if such a reminder were necessary, of the prior entente between the subject and the external world, the pertinence of which I have drawn attention to on many occasions. That is why the authors speak of a similarity between the state of wakefulness, as an intrinsic structure to the brain, and paradoxical sleep (Llinás and Paré, 1991, p. 524). In this connection they give the example of persons suffering from a bilateral retinal detachment, who none the less continue to have intact dream imagery during paradoxical sleep. "Moreover, neuropsychological investigations indicate that even if dreams may seem irrational by waking standards, they still reflect the cognitive abilities present in the waking state" (ibid. p. 525). The sensible data of the external world are, in turn, subject to cerebral aptitudes. According to Llinás "the model of the world emerging during ontogeny is governed by innate predispositions of the brain to categorize and integrate the sensory world in certain ways" (ibid., p. 526). The brain succeeds, he says, in recreating during sleep models that resemble those of the external world.

The closed system of the encephalon does not prevent it from being subject to external exigencies. Llinás indicates, on the other hand, that the brain is capable of engendering intrinsic activities by itself. These are *oscillatory* in nature. He is referring to the state of consciousness generated by the oscillatory connections between the *cortex* and the *thalamus*. This loop is unevenly bidirectional in favour of the cortico-thalamic fibres. The authors add, moreover, that the number of fibres projecting from the cortex to the thalamus is much higher than those projected towards it by afferent sensory entities.

According to Llinás and Paré, consciousness is self-generated by means of the oscillatory loop between the cortex and the thalamus. True, this consciousness is *innate*, but its emergence is due to the interactions between the brain and its external environment. They add that the functioning of the nervous system is not only determined by connections, but also by the membrane properties of their *constitutive* elements which are endowed with oscillatory movement. This also applies to the functioning of the thalamus. Of course, the thalamus fulfils its role as a relay between sensory input and the cortex; but it also generates its own electric activity in an *intrinsic* manner. These two thalamic functions are closely bound up with each other. This autorhythmicity is shared by other cerebral structures that are being discovered day by day. "Thus a large part of thalamocortical connectivity," they write, "is devoted to re-entrant or to reverberating activity . . . The insertion of neurons with intrinsic oscillatory capabilities into this complex synaptic network allows the brain to generate global oscillatory states which shape the computational events evoked by sensory stimuli" (ibid., p. 526). It follows that the different brain states, such as wakefulness or sleep, are particular examples of *variations* of the self-generated brain activity. "In other words, sensory cues gain their significance by virtue of triggering a pre-existing disposition of the brain to be active in a particular way" (ibid., p. 527). Only such a cerebral organization would be able to take account of complex activities such as wakefulness or paradoxical sleep.

The originality of Llinás' thesis lies in his ingenious synthesis of simple elements which, moreover, makes good sense. The first of these is the finding, derived from his own scientific approach, that the brain is a closed system. From there, he comes to the conclusion that *consciousness is an intrinsic property*. This completes the discrete upheaval that he has brought about in neuroscience, an upheaval that consists in discarding the main, and yet unexamined, idea that consciousness is an epiphenomenon of sensory information.

But what is consciousness and what does it correspond to on the neurophysi-ological plane? In response to the second part of the question, Llinás argues that the loop of thalam-cortical dialogue is a constant phenomenon both dur-ing wakefulness and during paradoxical sleep. These oscillations, the frequency of which reaches 40 Hz, fall to an insignificant level during slow sleep. We can therefore see the fundamental similarity that exists between wakefulness and paradoxical sleep. Their common point is consequently what is generally called the phenomenon of consciousness, which is altered during sleep. The presence of sensory input constitutes the difference between wakefulness and sleep. During the latter, this input remains *inhibited* in order to give free rein to the intrinsic data. Llinás remarks that during paradoxical sleep the energy of the encephalon puts itself more in the service of attention and memory. The brain therefore draws on its own resources, whose amplitude can exceed the bar of 40 Hz. This alteration is "reset" on waking thanks to the renewed pres-ence of sensory input. Consciousness, altered during sleep, then rediscovers its unshakeable ally in the form of the external world.

Llinás' neurophysiological theory on the alteration of consciousness during paradoxical sleep merely confirms dream theory in psychoanalysis. Freud estab-lishes a dynamic relation between the dream, which is the guardian of sleep, and the censorship which inhibits the action of the ego, as the guardian of the external world. The intrinsic functioning of the encephalon is another major element that may be regarded as consistent with the Freudian discovery.

Another original aspect of the research of Llinás and his collaborators lies in the discovery of the constancy of the radial frequencies of the encephalon at the level of 40 Hz, a phenomenon shared between the waking state and the sleep of dreaming. The constant loop of bidirectional exchanges between the thalamus and the cortex confirms, Llinás says, that it is indeed a question of the phenomenon of consciousness. According to him, it is thanks to this same phenomenon that the individual experiences his subjectivity. For Llinás (2002), "the issue of cognition is first and foremost an empirical problem, not a philosophical one" (p. 113).

If Llinás' research and theoretical articulations were confirmed and deepened by other neurological investigations, this would constitute a major advance in philosophical questions such as time and consciousness. Will the constancy of electrical frequencies in the cortico-thalamic loop give a new impulsion to the notion of time lived as *duration*? (See Bergson, 1965.) Would we be justified in conceiving the duration inherent to neurological functioning as a primordial condition of the access of living things to temporality?

According to Llinás, consciousness as the sense of oneself flows from the constant flux of exchanges between the thalamus and the cortical structures. If such an allegation proves to be exact, it would render obsolete, once and for all, the old metaphysical conception of consciousness as substance. From the great British neurologist, Charles Sherrington, to Antonio Damasio, including Eccles on the way, the trap of the substantialist notion of consciousness has proved constant. For the first time in the history of neurology, a scientist has dis-tanced himself from such a fallacious entity by affirming consciousness, not as a substantial ontological entity, but as the dynamic state of duration inherent to the experience of living things. It is also the first time that a neurologist has ceased to conceive of the mind as being an aggregate of the different cerebral structures.

Dream work and neurophysiology of sleep

After identifying the attempt to adventize as a function of oneiric activity, I have highlighted its internal logic, namely, its (visual) representability (*figurabilité*), its intrigue, its narrativity, its mise en abyme of the subject, and its linguistic devices. We are now going to study the mechanisms that govern what Freud calls the dream work (*Traumarbeit*). The common denominator of these mechanisms is that they give the dream an appearance of bizarreness or incongruity. We would like to know the reasons for these strange distortions at the psychoanalytic level as well as on the neurophysiological level. This will lead us to explore more closely the difference between sleep and dreaming, a difference that sometimes tends to become blurred in neurophysiology. As we shall see, this can even entail denying all the subjective content of the dream in favour of states of consciousness that are supposed to be pure products of neurological mechanisms.

What, in fact, is a dream? Is dreaming an act or a state? Does it proceed from an activity of cerebral neurons that is a matter of chance and inconsequential? Or does it proceed from a higher cerebral order? Is it proven that its triggering is due to the brainstem as the researches carried out in France by Michel Jouvet suggest? We know that the brainstem is the most archaic part of the brain, one, moreover, that is called reptilian. Should we conclude that the dream is also the product of this archaic state? Is it a regression or does it proceed from a high place of speech?

It was in 1953 that Kleitman and Aserinsky published their research on sleep. This immediately opened up the question of dreaming, since the two researchers had studied the relationship between the sleep state and the REM that are observed during this state. Electrooculargraphic (EOG) and electroencephalographic (EEG) recordings suggested there was a correlation between REM and a particular episode during sleep in which brain activity was in full swing, engendering a dream state at least equivalent in intensity to the waking state. The researchers decided to wake up the sleeping persons during the different stages of sleep in order to see which episodes were likely to engender dreaming. The results suggested that REM was the phase during which dreams occurred with all their bizarre and incongrous characteristics. Dreams that occurred during other so-called Non-REM phases seemed closer to the preoccupations and reflections of the waking state. There then appeared the existence of a specific phenomenon within the sleep state, namely the quasi-equivalence recorded between the intensity of brain activity during the REM phase and the phase favourable to dreaming, that is, during so-called paradoxical sleep.

Henceforth the laboratories of neurophysiology flourishing everywhere in the world began doing further research into paradoxical sleep (REM). The equivalence established between the latter and dreaming occupied the foreground to the detriment of studies into non-paradoxical sleep called NREM. Neurophysiologists also became interested in NREM to the point that the dominance of the theory equating dreaming with paradoxical sleep began to decline. McCarley and Hobson were among the most determined supporters of this equation. This current of thought is distinguished by its bitter opposition to Freudian theory. This can be seen from the famous article published by McCarley and Hobson (1977) on the *Project for a Scientific Psychology* (1950 [1895]), a manuscript Freud wrote in 1895 for the attention of Fliess and published in 1950 by Ernst Kris.

As McCarley (1998) writes, "The thesis of this chapter is that dreams are best regarded as transparent reflections of a brain state and that, contrary to Freud's view, dreams are not formed to disguise forbidden wishes" (p. 116). Henceforth the dream was principally assimilated to sleep, which is a specific brain state. This principal idea gave rise to the theory known as *activation-synthesis*. According to this theory, the brainstem (PGO activity), on the strength of its neuronal functioning during paradoxical sleep, sends stimuli of all kinds to the forebrain which, depending on its momentary psychic state and its past history, tries to integrate them by "disguising" them in a narrative register, often full of incongruous intrigues. Thanks to the PGO activity triggered in the brainstem, the retina is highly stimulated and produces REM. These movements are oriented towards the thalamus which, at the same time, receives stimulations from the brainstem. It is now that the thalamus, caught between the two types of activation (the brainstem and the retina), plays its role of relay by activating the higher-order areas of the cortex. The latter then tries to give form and consistency to the stimuli received by integrating them in the form of dreams. This process has two parts to it: the *activation* of the stimuli engendered by the reticular formation of the brainstem (PGO), on the one hand, and the *synthesis* by the higher brain of these stimulations (dreaming) on the other. This was what gave rise to the activation-synthesis hypothesis of McCarley and Hobson. These authors give an instructive example: "We know from the neuronal recordings in animals," McCarley writes, "that the vestibular complexes of the brain stem are highly activated during paradoxical sleep" (1998, p. 116). The author then gives the example of a dream that illustrates, in his view, vestibular sensations, that is, those linked to the internal ear responsible for the body's state of equilibrium. In *The Dreaming Brain*, Hobson (1988) writes:

> I was spinning, my body was spinning around. The circus performers put the bit in their horses and they spin around. The trapeze was spinning like that. Hands at my sides and yet there was nothing touching me. I was as nature made me and I was revolving at 45 rpm record (speed). Had a big hole in the center of my head. Spinning, spinning, and spinning. And at the same time, orbiting. Orbiting what, I don't know. I'd stop for a second, stop this orbit and spinning.

(p. 244)

McCarley (1998) comments on this dream. He writes:

> Of course, this dream and others in the series were not from circus performers astronauts who had such strong vestibular sensations as part of their daily experience, but rather were from college students who spent a large part of their time studying and who exercised in a more earthbound manner. The activation-synthesis model suggests that it is the presence of REM sleep activation of sensory systems, such as the vestibular system, that furnishes the basic sensory material, with the individual synthesizing this information in a way that best melds with his current state and past history.
>
> (p. 128)

Now we know that an experienced clinician could not fail to be worried by the occurrence of such a dream in a young student. Does it represent a failed attempt of adventization in order not to sink into psychosis? We know that dreams have a precursory role in relation to such tragic events.

These kind of precursory dreams anticipating psychic collapse may be considered as much as attempts to adventize an imminent danger as desperate appeals to the Other in order to benefit from his protection. I will take up *in extenso* two important observations by Médard Boss which illustrate brilliantly this kind of muted scream marking the extreme despondency of a person heading towards a fatal destiny but who tries, nonetheless, to find a way out. Boss (1957) writes:

> A woman of hardly 30 years dreamt, at a time when she still felt completely healthy, that she was afire in the stables. Around her, the fire, an ever larger crust of lava was forming. Half from the outside and half from the inside of her own body, she could see how the fire was slowly becoming choked by this crust. Suddenly she was entirely outside this fire and, as if possessed, she beat the fire with a club to break the crust and to let some air in. But the dreamer soon got tired and slowly she (the fire) became extinguished. Four days after this dream she began to suffer from acute schizophrenia. In the details of the dream the dreamer had exactly predicted the special course of her psychosis. She became rigid at first and, in effect, encrusted. Six weeks afterward she defended herself once more with all her might against the choking of her life's fire, until finally she became completely extinguished both spiritually and mentally. Now, for some years, she has been like a burnt-out crater.
>
> (p. 162)

I am now going to include another dream cited by Médard Boss:

> [. . .] a girl of twenty-five dreamt that she had cooked dinner for her family of five. She had just served it and she now called for parents and her sisters to dinner. Nobody replied. Only her voice returned as if it were an echo from a

deep cave. She found the sudden emptiness of the house uncanny. She rushed upstairs to look for her family. In the first bedroom, she could see her two sisters sitting on two beds. In spite of her impatient calls they remained in an unnaturally rigid position and did not even answer her. She went up to her sisters and wanted to shake them. Suddenly she noticed that they were stone statues. She escaped in horror and rushed into her mother's room. Her mother too had turned into stone and was sitting inertly in her arm chair staring into the air with glazed eyes. The dreamer escaped into room of her father. He stood in the middle of it. In her despair she rushed up to him and, desiring his protection, she threw her arms round his neck. But he too was made of stone and, to her utter horror, he turned into sand when she embraced him. She awaked in absolute terror, and was so stunned by the dream experience that she could not move for some minutes. This same horrible dream was dreamt by the patient on four successive occasions within a few days. At that time she was apparently the picture of mental and physical health. Her parents used to call her the sunshine of the whole family. Ten days after the fourth repetition of the dream, the patient was taken ill with an acute form of schizophrenia displaying severe catatonic symptoms. She felt into a state which was remarkably similar to the physical petrification of her family that she had dreamt about. She was now overpowered in waking life by behavior patterns that in her dreams she had merely observed in other persons.

(Boss, 1957, pp. 162–163)

The term activation-synthesis expresses it well. There are two parts to the dream and its neurophysiological mechanism: the (automatic) activity engendered by the brainstem (activation), and the meaning that the anterior brain bestows on it (synthesis). The problem that arises is the predominance of the mechanical activation produced in the brainstem by the reticular formation (the bulb, the pons, and the mid-brain) over meaning. The latter is merely an elaboration subject to the triggering mechanism of PGO activity. Even if the meaning in question has its roots in the dreamer's individual history, it is devoid of depth.

There are dreams that are, so to speak, inconsequential. They resist all efforts to understand them clinically. But they have nothing to do with the type of alarming dreams that I have borrowed from McCarley's book.

A man of about forty had the following dream. "I was watching television with my wife. We were no doubt in one of my childhood settings. The screen was small, but housed in a much bigger frame. We were watching a football match that was taking place in Spain. The match was over, but the players from each team were involved in a violent scene with the opposite team. I moved closer to the screen and saw, without being astonished, that a kind of mincemeat mixed with vegetables, or perhaps with fried rice, was coming out of it. In front of this mixture I placed a packet of food grain for guinea pigs. Then I woke up. I realize that I have forgotten one detail. In seeing the violent images on the screen, I expressed my indignation to my wife, saying ironically something like, 'You see how civilized they are, those people!'"

It is tempting to think that the dream imagery is the pictorial expression of the words reported at the end of the dream narrative. In this case, the mincemeat as well as the guinea pig food could all be allusions to the bestial aspect of the violent scene. Why does it occur in a childhood setting? What is the relation between the dream content and Spain? What do the words addressed to his wife allude to? Had he heard this said in reality, as Freud often supposes concerning words heard in dreams? We would be justified in supposing that the dream is without purpose, without consequence. Perhaps it is the failure of an attempt at adventization.

McCarley and Hobson's activation-synthesis hypothesis rejects the existence of a latent content in dreams. In their view the meaning given to the activation in question is literal and therefore transparent. The synthesis of the forebrain is considered on each occasion to be dependent on the pattern of movement produced by the reticular formation of the brainstem. The similarity in question is the concordance between the endogenous activity of the brain and its activity in response to the external world, that is to say during the waking state. In spite of the paralyzed muscular tonus, the neurons receive the same patterns of movement as during wakefulness. Consequently, the higher part of the brain takes them as such, that is, as real movements investing the muscular part. Let me cite Hobson again:

> The pattern of discharge [of the visual neurons] is pulsatile, just as it is in response to external visual stimuli. Dreams are characterized by a sense of continuous movement, and brain neurons concerned with movement fire intensely during REM sleep. Their pattern of discharge is pulsatile, just as it is when movement is generated in the waking state.
>
> (Hobson, 1988, p. 171)

Hobson notes here the similarity at the neurological level between the modes of triggering of perception and the dream image. As he sees it, the brainstem triggers patterns of sensation and/or movement during paradoxical sleep in the same way as during the waking state. What the accomplishment of the act is lacking is, in effect, the muscular system that is paralyzed by the postural atonia characterizing the dream state. According to the activation-synthesis model, the dream is a *random synthesis* that the higher brain carries out in relation to the mass of endogenous perceptions and at the level of the neurons of the reticular formation of the brainstem. When the patterns of movement/sensation are triggered by the latter, the cortex reacts by drawing on what is available, namely, its memory. To this he adds meaning and significance,[4] even though the patterns are originally only pure excitations of neurons located in the reticular formation. Clinical experience, however, formally belies the hypothesis of random sensation. The overdetermination that one finds in analytic work between the dream and the subjective history of the individual weakens the thesis that dream activity is at the mercy of stimuli devoid of any psychic aim. The following examples highlight the overdetermination that is involved.

Mathieu is a man approaching his sixties. He came to analysis on account of depression following his divorce. Mathieu is his gallicized name, chosen at the time of his naturalization. On the social, family and professional level, he is a prime example of perfect integration within French society. The only tie he still has with his former family is his son, the issue of a previous marriage in his country of origin which had ended in divorce. Having broken off all contact with his country, he doesn't even have any opportunities to practise his mother language. And yet he is haunted by a hypnagogic image as he is going to sleep: his father appears before him almost every evening in an immobile posture against a blurred or imprecise background. It is true that he learnt of his father's death through a chance encounter in the streets of Paris. The systematic occurrence of this image only ceased following a series of dreams over several months during which the patient was constantly tempted to put an end to his analysis, especially as the feelings of grief that had led him into it now seemed to have ceased. It would be tedious to enter into the details of this analytic work that was difficult to see through to its end. I will simply report here a few of his dreams that occurred during the last months of his analysis.

"I dreamt that I had a black child. He was standing at my side. I seemed to be very tall in comparison with him. At a certain moment, I took his hand in mine. Seeing his hand all black, I felt full of tenderness towards him. There were many other things in the dream that I have forgotten. I then found myself with all my children in a boat that no doubt ran a shuttle service between the place where we were living and a far-off country which, paradoxically, seemed to me to be very close. My children hid in the hull of the boat. Suddenly, I was terrified by the idea that they had embarked to go to this country forever. We got out of the boat quickly before it was too far away from our shores."

The same night he had the following dream: "I was in a cemetery, most probably in Russia. I was badly dressed and unshaven. No one spoke to me; a lot of people were mourning. Everyone was getting ready to enter the funeral parlour. My eyes fell on the Russion inscription that was on the frontispiece. I tried to read it. I made out the word: *toniev*.

"Both dreams occurred during a night when my sleep was interrupted by an obsessive thought. I was trying to remember the name of one of my students. I knew that it began with the letter "m", but I only remembered it when I woke up."

The evening before, he had had a discussion with one of his friends who had just arrived from the United States. This friend had made some ambiguous remarks about black people. To avoid having a dispute with him, he refrained from replying. This must have reminded him of several memories which are present in the dream. The far-off/nearby country, it should be added, is no other than his country of origin. The boat is an allusion to the time when his situation of residence in France had proved problematic. At the same period, he had fallen passionately in love with a young woman who finally preferred her ex-lover to him. "She was," he said, "a woman of exceptional beauty and tenderness who complained about her small waist." To this he associated his tallness in the dream where he felt proud of having become black through the child in the dream.

In addition there was what he called the "Russian whiteness" of the second dream. Russia was for him a country of high psychic value. He had made a short stay there during a scientific expedition. The trip had been shortened even further because he had fallen ill, and only recovered once he was back in France. It was now that he was wondering about the reason for this sudden illness. The country had reminded him strongly of his own, on account, he said, of the corruption.

Becoming black-skinned was a form of revenge, both with regard to his friend from America with his discourse about Blacks and with regard to the young woman who was singled out by the whiteness of her skin. The recollection of this disappointing love affair had a paradigmatic function for him of what may be called an unhappy love affair, mixed, furthermore, with the pain of exile. The young woman was called Francine. As he was in danger at the time of having to leave France, the name had acquired for him its full literal sense. His immense grief, reactualized in the dream, in relation to the young woman, was now mixed up with the torment that had prompted him to undertake an analysis. His sadness had stayed with him throughout the whole night of the dream. Indeed, he could not stop thinking about this student who not only evoked for him his dashed love affair, but whose first name, Morgane, had a particular resonance for him. It evoked the ultimate sadness that is death. We can easily understand now why her first name only came back to him once he had woken up. The dream attempted to repress the idea of death by transforming it into sorrow for a lost love, but linked to a country that had welcomed him. The signifiers linked to a given country often possess a symbolic connotation in dreams, which is linked to the name-of-the-father. The inscription on the frontispiece of the funeral parlour[5] was, in fact, simply the transformation of the first name of the deceased father.

To return now to our discussion. In the activation-synthesis theory, the dream is only a by-product of sleep. The brainstem triggers sensory and motor patterns which activate the higher brain in turn. If even we accepted that dreams are as many syntheses forming in response to neuronal stimuli from the reticular part of the brainstem, we still need to know why the activation-synthesis theory ascribes them a transparent content without consequence for the subject's psychic life. The production of the dream may, after all, be supported by a somatic function and, what's more, without being devoid of great significance. Arguing the contrary would be more redolent of an ideological position with regard to Freudian theory than of results stemming from scientific work.

The continuity between sleep and dreams, as suggested by the activation-synthesis theory, is in opposition to Freudian theory. According to the latter, sleep and dreams stand in a relationship of conflictual complementarity. The dream is the guardian of sleep only to the extent that it is menaced by the latter. The temporal and functional proximity between paradoxical sleep and waking confirms this. Paradoxical sleep means that we only dream when we are almost awake. In this connection, Freud (1900) writes:

> A very attractive conjecture put forward by Goblot suggested, no doubt, by the riddle of Maury's guillotine dream. He seeks to show that a dream occupies

no more than the transition period between sleeping and waking. The process of awakening takes a certain amount of time, and during that time the dream occurs. We imagine that the final dream-image was so powerful that it compelled us to wake; whereas in fact it was only so powerful because at that moment we were already on the point of waking. 'Un rêve c'est un rêveil qui commence' ['A dream is an awakening that is beginning'].

(p. 575)

Sleep and the waking state

The radical alterity that Freud assigns to the unconscious is such that it justifies his constant concern to distinguish it from the preconscious/conscious register. This differentiation assumes a topological configuration in Freudian theory. It is this "topographical localization" in Freud's work, albeit theoretical and mythical in nature, that leads us to discover the affinity between the psychical apparatus of *The Interpretation of Dreams* and Llinás' thesis of the cortico-thalamic loops.

It is a question of the rapprochement between the psychical apparatus in *The Interpretation of Dreams* – the purpose of which is to demonstrate the mechanism by which unconscious ideas make their *return* to the oneiric stage – and the thesis of the oscillations of the cortical structures and the thalamus which link up external sensory input with these same structures.

The dominant thesis in the neurophysiology of sleep tends to attribute the functioning of paradoxical sleep to what is known as *muscular atonia,* which is triggered by the brainstem. The latter is thought to be equipped with a mechanical system (*on/off*) whose functioning is activated by chemical transmitters. This concerns, according to the activation-synthesis model, three nuclei situated on the brainstem, namely the tegmental tract (a dense network of fibres on the upper part of the brainstem which is responsible, it is supposed, for the triggering of paradoxical sleep where acetylcholine is secreted), the dorsal raphe nucleus, and finally the locus coeruleus which puts an end to paradoxical sleep by secreting aminergic acid.

But as long as it confines itself to the purely mechanical level, the activation-synthesis model will lack a basis for explaining the dream/waking cycle. This can be seen from sleep disorders which often confront the neurologist with his scientific indigence. Without taking subjective factors into account, neurophysiological investigations in the domain of sleep risk leading us into clinical aporia by condemning us to mechanical explanations.

For Llinás, just like Freud, the question of sleep is linked more to the alteration of the state of consciousness. This suggests that the subjective dimension plays an important role in the process of sleep. It is certainly not a matter of an all-powerful ego that decides deliberately to sink into the altered state that is sleep or to emerge from it; but rather of a complex dynamic in which the *sleeper* mind becomes a *sleeping* subject. This is what Llinás calls an *altered state of consciousness.* For his part, Freud calls it the *wish* to sleep. The denomination *paradoxical sleep* expresses it well. During it we are in an oscillating state between dreaming and waking. During sleep, the state of consciousness is already, thanks to the sleeping subject, oriented towards the loops of neuronal activities in favour of structures that are more cortical than thalamic, leaving behind the sensory input whose activity remains imminent at each moment during paradoxical sleep. This brings us back to the maxim according to which it is always the animal as such that is sleeping or waking up, and not its system of neurotransmitters. According to Llinás, it is, quite to the contrary, altered consciousness that is determinant in the onset of sleep and not muscular atonia, as the dominant explanation would have it.

Certainly, the very structure of dream activity is based on the asymptotic relation between the dreamer and the dreaming subject, governed by the division of the subject which is intrinsically bound up with the Other; but it is important to add that it remains dependent on the oscillating and uncertain relationship between sleep and the waking state. The progressive process of slipping from the waking state towards somnolence in the latency characteristic of falling asleep, as well as that which brings the subject back to wakefulness, remains highly complex. Both are characterized by an often precarious equilibrium between the phantasy life of the subject and the external conditions of his everyday life of which day residues are a part. It is true that the latter nourish nocturnal life, but only to the extent that they do not conform to the regulation brought about by the subject's phantasy life. They are as many failures of this regulation. And it is in this respect that they disturb the sleeper's nocturnal world. Hence the attempt of oneiric adventization which highlights them by accentuating their point of failure. Conflict breaks out, then, between the wish to sleep and the need to adventize the obstacles put in the way of phantasy life, whose mission is especially to regulate the subject's everyday life. Here lies the real cause of sleeping problems.

The subject's phantasy life obviously does not mean that he is cut off from the external world. Freud has taught us to appreciate the regulatory role of phantasies. Far from being an illusory production, phantasies play a prominent role in the stability and organization of what is known as psychic *reality*. Phantasies are operative both upstream and downstream in relation to the world of dreams. They participate as much in daytime reveries as in the formation of dreams. They are equally present in the secondary elaboration that follows on awakening, in the form of a defensive distortion and a revised narrative of the dream. In other words, once he has woken up, the subject tries to return to his usual world, hidden as it is behind phantasy constructions. Without being divorced from the external world, psychic reality not only conforms with it but also plays an active part in its maintenance and constancy. And so we are in a position to speak of the being-in-the-world of man not only as the subject who constitutes the lived temporality of the world, but also as one who is subject to its fluctuations and uncertainties.

Employing the language of computer processing, Llinás describes waking up, that is, the subject's return to his waking life, as a resetting of the sensory world. This restoring of order can obviously not be reduced to the putting back to zero of a world that has disappeared for a while before reappearing again. It is always the being-in-the-world of a subject who is divided within himself and who shares this condition with his fellow human beings.

Notes

1 PGO activity = ponto-geniculo-occipital waves, described later in this chapter.
2 See below the section on play and the dream work.
3 See https://sommeil.univ-lyon1.fr/articles/savenir/electricite/electricite.php
4 Hobson and McCarley agree on the principle that the dream is not devoid of meaning. But, for these authors, such meaning is transparent and recourse to interpretation is not required. And so they put the emphasis on endogenous perceptions, triggered by the brainstem in a random way and the random synthesis that the anterior brain carries out in dream formation (see McCarley, 1998).
5 Cf. *morgue*, Morgane

8

DREAMING, AN INTIMATE CHARACTERISTIC OF THE MIND

The mechanisms of sleep, a new approach

Are we justified in tracing oneiric activity back to the triggering of the mechanisms specific to the brainstem? Is dream imagery linked to rapid eye movement in correlation with PGO activity or is it rather the forebrain, with its multiple structures, that is responsible for dreaming?

What is dreaming, and what about sleep? Does the division of sleep into REM and NREM correspond to the need to distinguish dreaming and sleeping? Characterizing dreaming as paradoxical sleep means that dreaming and sleeping are by definition in opposition to each other. Sleep is a time of rest whereas dreaming attests, on the contrary, to intense brain activity agitating almost the whole of the encephalon. The coincidence of dreaming and sleep cannot, therefore, be simply a matter of a paradox.

We know that for Freud paradox is an integral part of the drive. Being a constant force, the drive tends, according to the so-called law of inertia, to return to the zero level of excitation. Can we ascribe to dreaming, to this high cerebral activity during sleep, the same paradoxical function that consists in reducing to the minimum the tension that nonetheless increases owing to its occurrence?

The dream is the privileged locus of the drive impulses to the extent that it is the paradigm of what characterizes the drive. Evidence of this can be seen in the loss (i.e. missing nature) of the object that proves constant in dreams. The impossibility of reaching the drive aim is another feature of it. This impossibility manifests itself through the sort of circumventing of the object characteristic of dreams. The lure with which the object of the drive is cathected fulfils its function marvellously during oneiric activity. In its closed circuit, the drive begins its trajectory from its somatic source called the erogenous zone and reaches its final point which Lacan designates by the English term *goal*. The loop or the closed circuit of the drive engenders its famous back-and forth-movement around its object. Each time around the circuit, the goal of satisfaction is not attained, but missed; hence its forever renewed departure. This closed circuit designates what Freud calls the partial drive.

In this circuit there is another constitutive element to be determined. The loop constantly closes itself again around what engenders its movement, namely, the object. The latter is the paradigm of what Freud calls the lost object, which has to be refound over and over again. We know that this object is not lost, even if it has to be found again. This sensation of reunion in the drive is named very judiciously by Freud as *Drang*. It signifies the force, the urgent pressure, what presses and seeks to attain its aim without delay, what can no longer wait, that is to say, expectation *par excellence*. Owing to its pressing character, expectation is brought back to the present time and the optative is transformed into the indicative.

Defining the drive as the junction between somatic and psychic, Freud underlines the incompleteness by which it is marked. This constant trait of the drive not only bestows on it its irremediable partial character, but also the uncertainty regarding its destiny (see Freud, 1915b). The drive is by essence partial. This fundamental characteristic justifies its use in the plural. The unification of the drives is henceforth marked with the stamp of the impossible. This impossibility is bound up with their incompletion and their partial character. It is probably from the same drive configuration that the fragmentary or piecemeal aspect of the dream emanates.

If the dream is supposed to be a drive, how can its subjection to the law of inertia be proven? How could a neurological flurry like oneiric activity accomplish the drive-related task of reducing the tension in question to its lowest level with a view to extinguishing the excitation that characterizes it?

The advocates of the activation-synthesis theory portray sleep as chaotic impulsions due to the specific mechanisms of the brainstem. It would not be erroneous to see in these impulsions the work of the partial drives. What is more piecemeal or fragmentary than the dream scene, which manifests nothing but instability and turmoil? It may be assumed that these impulses are in the immediate proximity of the organic instinctual impulses of a brain that is prey to their agitation. This chaotic *activation* is matched by oneiric activity as *synthesis*.

What is the aim, then, of oneiric activity in the face of such activation? First of all, the dream only occurs, in keeping with the paradoxical function of the drive, to reduce the excitation in question to its lowest level. It is indeed the guardian of sleep and has the task of maintaining its function of procuring rest and relaxation. The sleeper's uppermost wish, Freud says, is to sleep. Such is the paradox of the dream: an intense activity aimed at bringing the excitation back to its lowest level. It is by means of such a function that oneiric activity attempts to bring about the synthesis of the partial drives. This coincides with elements of the activation-synthesis theory. But such a synthesis is by definition impossible. The drive is by essence, as I have said, partial. The dream always remains fragmentary. The attempt at synthesis takes place none the less. But it does not totalize the drives. Rather, it results from their own convergence.

Vicissitude is inherent in the drives. The turning round into its contrary, regression, introversion, the passage from activity to passivity, the turning round upon the self and sublimation are all drive vicissitudes. The components of dream logic, that is, narration, intrigue, the mise en abyme of the subject, linguistic devices and their putting into image form must also be counted among these vicissitudes. It is in this way that, when the right moment comes, drives can attempt to unite in order to gain access to adventization. Such an attempt can occur at any moment whatsoever during sleep, the most favourable being, of course, paradoxical sleep. This would bring an end to the quarrel concerning the occurrence of dreams during REM or outside it.

It could be conceded, with McCarley and Hobson, that it is primarily the brainstem that triggers paradoxical sleep. The brainstem, as we know, is the most archaic part of the human brain. Hence the rapprochement that some authors have ventured to make between the activation of this part of the brain and the primary process in Freudian theory. The primary process involves the partial drives but, generally speaking, they do not remain at this stage.

However, restricting the triggering of paradoxical sleep to the brainstem would be equivalent to localizing the drives, which would take us back a century to locationalist neurology. The vicissitudes of the drives are in keeping with the brain's plasticity. Perhaps one day it will be possible to establish the cartography of the brain on the basis of the drives and their mode of functioning.

Should the triggering function of paradoxical sleep be attributed to the brainstem or to the forebrain? We can see that with regard to the nature of the drives, their lability and their vicissitudes, such a debate can lose its reason for existing.

Dream imagery is one of the distinctive signs of the vicissitudes of the drives. Remember that Freud traces the drive back to the joy of seeing (*Schaulust*). At the same time he attributes the drives with the role of ideational representative linking the somatic and the psychic. In this conjunction, imagery (REM) plays a capital role during paradoxical sleep. We will see that this is explainable by a specific kind of seeing that is characteristic of the dream.

The activation of the partial drives is matched by their attempt to achieve a form of synthesis characteristic of dreams. This synthesis, which I have called adventization, is not apparent in the manifest content of the dream.

Accordingly, adventization may be considered as *the reverse side of the drive* inasmuch as it results from the convergence of the partial drives, which together succeed in finding an aim and in assuming another destiny. Adventization is dependent on dream logic, which in turn is part of the vicissitudes of the drives themselves. By virtue of its composition, adventization reflects the dispute and the conflict that set the drives in opposition to each other. Adventization, as drive convergence, remains dependent on this kind of tugging. It would be more exact, I know, to speak of an *attempt* at adventization. Indeed, before a dream is formed, it must go through numerous obstacles and multiple processes.

The question concerning the object of the drives is of such complexity that we will have to come back to it later with ample clinical evidence.

The cognitivist hypotheses

The Finnish cognitivist philosopher, Antti Revonsuo (2000), has proposed a coherent theory on the function and usefulness of dreaming. The author is critical of the activation-synthesis theory of dreaming which "claims that dreaming is a random by-product of REM sleep physiology" (Revonsuo, 2000, p. 877). The main question he asks is why we dream. Does dreaming have a purpose, a function? "The leading neurocognitive theories," he writes, "seem to have given up the hope of identifying any useful function for dreaming at all" (ibid). Revonsuo advocates making a clear distinction between dreaming and sleep. The latter is a physiological state, whereas dreaming "refers to the subjective conscious experiences we have during sleep" (p. 878). He recalls that recent research has clearly shown that dreaming can occur outside paradoxical sleep and that the latter can also turn out to be without any dream activity. He raises objections to the Hobsonian theory which stresses the random aspects of dream imagery. It is important to remember that, according to this theory, PGO waves are produced during sleep by the pontine part of the brainstem which, in turn, activates the forebrain. The forebrain attempts to make sense of this random activation by making a specific synthesis of it at the level of dream imagery, in keeping with these same stimulations. The images are borrowed from memory in which waking residues have an important place. "The theory delivers no answer," he adds, "to the question why the brain should generate any images at all during REM sleep; it is simply assumed to be an automatic process. The narrative content of dreams remains unexplained as well" (p. 879).

In his search for a plausible function for dreaming, the author refers to several theories that are worthy of mention. He cites the theory presented by Crick and Mitchinson (1995) according to which memory is compared to simple models of associative nets. When such a net gets overloaded, it produces different combinations of stored associations. In order to avoid overloading, a reverse process is set in motion. Its inputs and outputs deviate from the normal process and produce random phenomena. Accordingly, the associative nets are weakened, which would explain the disconnected aspect of dreams. Expanding on the same theory, David Foulkes, one of the major figures of the neurophysiology of sleep, situates dreaming at the level of the random activity of semantic memory occurring episodically during paradoxical sleep. Foulkes, however, establishes a subtle distinction between dream content and dreaming as a process. Dream content links up in a more or less coherent way the subjective events unfolding in "an analogous world" with the subject's active participation. But Foulkes does not assign any function to dreaming.

Revonsuo then turns his interest to the recent theory of Solms (1997) which defends the Freudian principle that the role of dreaming is to protect sleep. During sleep, internal stimuli activate the "curiosity-interest-expectancy circuits" of the brain. Inhibitory mechanisms in the brain are activated to prevent "appetitive interest" from invading the motor system.[1] This is when a regressive process begins giving rise to hallucinatory images.[2] For Solms, the brain mistakes these for real perceptions. Revonsuo notes that, like his colleagues, Solms does not attribute any specific function to the content of dreams.

Let us pursue the debate on the function of dreaming further with other cognitivist authors. For Owen Flanagan (1995), dreams are only a "collateral" effect of the brain during sleep and in this respect they are devoid of interest at the biological level. In his view, dreams contribute nothing to the waking period of brain activity because the dream thoughts are not worth remembering by the brain. For Antrobus, too, given the absence of sensory information and motor commands during paradoxical sleep, the associative context of the brain has no importance. The act of dreaming has survived the uncertainties of evolution insofar, precisely, as it had no maladaptive consequences for the animal's organization.

Theory of the phylogenetic utility of dreaming

We are now going to return to Revonsuo's question about the possibility of finding a specific functioning for dreaming. Although it does not play a dominant role in neurophysiological theories, the question raised by Revonsuo is of increasing pertinence for the psychology of dreams.

Does dreaming help the subject to adapt better to his environment? Does it play a part in the individual's psychic equilibrium? Revonsuo traces his inquiry back to Jung's theory which assigns dreaming an eminent role in the psychic equilibrium of the individual (see Jung, 2010). He also mentions Adler (1925) who believed that dreaming had a problem-solving function in the subject's life. The experiments conducted with the aim of establishing whether dreaming in fact resolves these kinds of problems do not seem to be conclusive (see Dement, 1972; Montangero, 1993; Cartwright, 1999). Let it be said in passing that this form of investigation involves a fundamental misappreciation of dream activity regarding the symbolic dimension of desire, which finds itself reduced to its simplest expression of psychological motivation.

Revonsuo also mentions Kramer's (1993) theory, called the selective mood regulatory function of dreaming. Noting the surge of emotions in dreams, Kramer thinks that the function of dreams is simply to contain the emotive charge thereof. If the attempt in question is successful, the dream will not enter awareness or memory and will continue to play its role as guardian of sleep; but if the subject retains conscious memories of it, it will indicate the failure of the dream process. This theory seems, once again, to neglect the symbolic dimension, even though it constitutes the corner stone of the mechanism of adventization. This negligence is shared by other cognitive researchers.

To return now to the theory put forward by Revonsuo himself concerning the specific function of dreaming. The author refutes, as I have said, the hypothesis of the random mechanism of dreaming. The organization of dream content is so elaborate, he says, that it could in no way be dependent on such a mechanism. It involves "all the sensory modalities" of the subject. Referring to the experiments of researchers such as Foulkes, Strauch and Meier, Revonsuo notes that the intensity of these modalities is comparable to that of waking life. The brain creates a world during sleep that is parallel to that of the previous day. In this dream world, we are "surrounded by a visuo-spatial world of objects, people, and animals, participating in a multitude of events and social interactions with other dream characters" (Revonsuo, 2000, p. 883). If dreams were just nervous noise, there would never have been such a complex organization. "True noise in the brain," he writes, "is produced in connection with an aura of migraine for example" (ibid.) Revonsuo refers to the experiments of Wilder Penfield, the famous Canadian neurosurgeon, concerning the electrical stimulation of the temporal cortex of his patients. We know that this kind of stimulation produced vivid and realistic perceptual "flashbacks". Revonsuo cites the author of these experiences who noted that such a "mechanism is capable of bringing back a strip of past experience in complete detail without any of the fanciful elaborations that occur in a man's dreaming" (ibid.).

According to Revonsuo, dreaming produces an organized model of the world. It is an organized simulation of the perceptual world. In other words, it is a "virtual reality". But the dream world is far from a repetition of our waking life. The studies carried out in this connection show this clearly. "Hartmann (1998), he adds, "describes two studies in which it was shown that even subjects who spend several hours daily reading, writing, or calculating, virtually never dream about these activities" (ibid.). And so Revonsuo draws his second conclusion: dreaming is not a systematic but rather a selective simulation of the world.

We are now in a position to formulate a certain number of observations concerning Revonsuo's theory. First, the perceptual equivalence that he establishes between the real world and the "virtual" world of dreaming is subject to caution.

Neither the imagination in general nor dream imagery in particular correspond so simply to the sensory world. This simplistic conception should first be rectified by stating that perception is neither a passive act on the part of a perceiving subject nor the copy or decal of external reality. The tuning between the subject and the world is such that we do not need to turn it into a question of correspondence. We know the metaphysical definition *veritas est adaequatio intellectus ad rem*, to which we referred in relation to the notion of the mind as a *tabula rasa,* which conceives of the mind as a locus of *re*-presentations.[3] Indeed, what Sartre calls the illusion of immanence dominates cognitivist research.[4]

The human perceptual world is dependent on language. It is language that gives meaning and consistency to imagery. We still have much to learn from dreams. We can hear in them the acoustic vacillation of words being transformed into image form. When a young girl has a veiled (blurred) dream about her father's marriage with her stepmother, what we hear her speaking about, in fact, is her own desire to be a bride in her veil attending the ceremony. This process of mise en abyme of the subject – since the dreaming subject is at once a witness and a participant – could not occur without recourse to language. The dream is not made to be described pictorially. It is made to be heard. To understand dreams, it would almost be necessary to recite them out loud like poetry; then we would hear their linguistic devices.

Language is intrinsically permeated by ambiguity, which is an integral part of our desiring dimension. It is a particular kind of ambiguity which, thanks to the richness of language, bestows on the dream its high expressive value.

A married man, who was getting a divorce, met his wife in a dream at the intersection of two little alleyways. After exchanging a few words on the subject of the forthcoming sale of their house, they said to each other, apparently with regret: "*on s'est bien manqué*". Given the extreme ambiguity of the words, it is difficult to know whether *ils se sont manqué* in the sense that they each missed each other or whether they realized that they had passed each other by, like boats in the night, without there being any real encounter. Indeed, the sale of their jointly-owned house embodied the extreme hesitation that each of them manifested in their feelings towards the other. They had started living together out of the need to conform to what they thought society required of them. The woman got pregnant but suffered an unexpected miscarriage towards the fifth month of the pregnancy. The conformity of their life was then substituted by a sense of indifference which barely hid their pain at having lost the child. Below is another example illustrating the inherent ambiguity of dreams.

A man in love with a woman learnt from her that one of his friends had been courting her. That night, he had the following dream.

"I was sitting, with the friend in question, on a bench in front of a pretty spring landscape. Behind us was the public W.C., where there were constant comings and goings. I said to my friend, pointing to the landscape, 'Just look how romantic it is'."

The dream no doubt has a characteristic of closure, an attempt to put an end to the dreamer's torments of love. Indeed, for some time now the sessions of analysis had been revolving around his unfruitful passion. Drawing on the manifest content of the dream, he took it as the expression of what opposed his romantic feelings towards his loved one and those of his friend, which he described as cheaply erotic. But things proved to be more complex than that. The signifier W.C. represented the name of the woman in question, save for the fact that her first initial was doubled. This doubling stood in, at the same time, for the "double use" of the woman as an object for the two men. We are no doubt in the presence here of the fall of the Thing (*das Ding*) that is the loved one to the rank of an object; what's more, an anal object, placed at the disposal of the public (see Freud, 1912).

Having given these two examples illustrating the ambiguity of dreams, we will now return to the critique of Revonsuo's theory.

The other observation that needs to be made in regard of Revonsuo's theory of dreaming concerns his conclusion about the selection of the lived events for the dream scenario. Why do the events of our daily lives not all find a way into our nocturnal activity of sleep? If adventization has a meaning, and if it is bound up with the symbolic, it would not be difficult to understand that the occurrence of the lived events during the dream process depends essentially on our subjective possibility of elevating them to the rank of the symbolic. I can only make brief mention here of a transference neurosis[5] that occurred at the end of the analysis of a woman in her forties. For several months she had been overwhelmed by this inconsolable passion for her analyst and had been unable, for days and weeks on end, to stop thinking about him. She said that she was in a real state of distress owing to the fact that she could not put her analyst out of her thoughts. Well, this state of affairs only stopped when she was able to dream about it. Her immense grief came to an end because her passionate love, implying the loss that was inherent to it, had been set on the path of adventization. From thereon, her wounded feelings could evolve towards a state of greater serenity and calm. Such is the selectivity of events in dream activity. It is also worth citing in this connection the unbearable tragedy of a father whose little girl, aged eight, was run over by a lorry backing up into their drive when she was on her bike accompanied by her elder brother. For months, during his sessions of analysis, he was overcome by interminable floods of tears which simply intensified increasingly his inconsolable pain. The first words he uttered, thereby interrupting this immense grief, came, in fact, in the form of telling a dream. Now, however, it was his daughter, lost forever, who reappeared in tears. This was the beginning of the first signs of the adventization of the drama, a process that had to continue for a long time before this heart-broken man, who, in addition, had to cope with the grief of his entire family, could find some semblance of relief.

Let us return directly to Revonsuo and pursue the elaboration of his theory. He is surprised by the massive occurrence of "negative elements" in dream content. He cites a study on dream content conducted on a thousand students affirming that out of a total of 700 emotions expressed 80 per cent contained negative emotions, such as "apprehension, sadness, anger, and confusion". Other studies cited by Revonsuo confirm the same results. It is not my intention to question the pertinence of such studies. I am simply following the author's line of reasoning. He further comments on the frequent absence of a "happy end" in dreams and notes the important place of aggression in dreams. He concludes that not only is dreaming a selective simulation of real events but that one of its major characteristics is related to threatening and aggressive situations of which the dreamer is the victim.

Revonsuo thinks that these negative elements reflect in many ways the life of prehistoric man, who was subject to threats both from nature and from his fellow-men. "In the ancestral human environment, however, intergroup aggression and the violent competition over access to valuable resources and territories is likely," he writes, "to have been a common occurrence" (2000, p. 885). This threatening situation was coupled, according to the author, with the proximity of life in nature where confrontation with wild animals was common. Even today, wild animals still populate the dreams of children even though there is nothing in modern life to justify it. "If dreams are naturally biased towards simulating," he writes, "then we should expect that these biases are strongest early in life, when the brain has not yet had the chance to adjust the biases in order to better fit the actual environment" (ibid.). This is why, in Revonsuo's view, we notice the presence of wild animals in the dreams of children. He then cites statistics, figures and curves to prove the importance of such an occurrence depending on the age of the child.

What we are witnessing here is the death of the symbolic. We do not know what Revonsuo would say about the presence of wild animals in the myths, fairy-tales and legends of cultures throughout the world. In his cognitivist approach the aspect of the enigmatic questioning that man of today (the child of yesterday) finds in the gaze of a wild animal – an animal, which, by virtue of its strangeness, reminds him of his own subjectivity – thus disappears. What place is more suited to the appearance of this uncanniness than the *other scene* that is the dream (*anderer Schauplatz*). On the subject of the lack of language in animals, Lacan (1967) said that,

> this 'keeping quiet' remains heavy with an enigma which, for such a long time, made the presence of the animal world heavy. We no longer have any trace of it except in phobia. But let us remember that for a long time the gods were put there.
>
> (Unpublished seminar, translated for this volume)

The animal's gaze is highly intriguing for the child because it conveys to him at once its proximity and difference with the other gaze which governs relations between human beings. Human relations are the privileged locus of desire, linking men to each other. Is the animal endowed with the same gaze conveying desire? Gaze governs the affairs of men so well that they cannot help turning away from it as soon as it ceases to serve them as a safe site of desire. It is because it haunts us as a limit to our relations with the real that we find the dead person's gaze unbearable. The darkness hidden in this gaze threatens us in more than one respect. It is connected with the traversing of a mirror which we had been able to avoid up until then, thanks to the desire that we had projected into it through a specular image. This image is maintained as long as it is supported by the presence of the Other – the very same presence which founders in the eyes of the dead person where our gaze proves to be a departure without return or horizon.

The animal's gaze highlights a fundamental mismatching which intrigues us precisely to the extent that it is not unreceptive to our own, giving us the *hope* of finding in it the signs of a desire which, instead of coming into being – for lack of language – is transformed into a privileged site of projections. It is no doubt true that it does question and interest the child, projection being his means of expressing play, but also his turmoil and fright.

Revonsuo's theory assigns a specific function to dreaming. In the "selective situation of dreams", the aim consists, according to him, of "repeating a threatening ancestral situation by simulating it in the current circumstances of the subject". Dreams, then, rehearse these threatening events so that the individual is better prepared to avoid them in reality. "I conclude," Revonsuo (2000) writes,

> that rehearsing threat-avoidance skills in the simulated environment of dreams is likely to lead to improved performance in real threat-avoidance situations in exactly the same way as mental training and implicit learning have been shown to lead to improved performance in a wide variety of tasks. It is not necessary to remember the simulated threats explicitly, for the

purpose of the simulations is to rehearse skills, and such rehearsal results in faster and improved skills rather than a set of explicitly accessible memories.

(p. 891)

This is how Revonsuo resolves the problem of the forgetting of dreams. This evolutionist theory claims that dreaming is a phylogenetic activity of the brain. From this perspective, it repeats ancestral experience with a view to improving the avoidance behaviours of man when faced with danger.

An argument that could be set against this theory consists in noting the ambiguity of the dreamer when faced with dangers. If we are supposed to improve our performance skills by repeating situations of danger in our dreams, how then are we to explain our ambivalence concerning them? Why don't we simply eliminate these dangers rather like those pre-adolescents who, in their ego-related dreams, are led to fight, with implacable mastery, every imaginary enemy? Is it not because these animals represent a part of ourselves? And is it not because the real violence is our confrontation with the real, that is, with our impossibility to adventize it? We can see, then, the tussle of the contradictory forces that inhabit us when the process of adventization is underway. This conflict resides, as we have seen, at the heart of the drive. It cannot fail, therefore, to be at its height during the oneiric drive which provides it with both a scene and intrigue.

We can understand now why animals occupy such a prominent place in all the cultures of the world. They are in effect the privileged link in the man's questioning of himself. The "silent subjectivity" of animals allows the child to experience intense pleasure in their company, as if he or she was able in this way to tame what, by definition, escapes him or her, namely, the opacity inherent to his or her subjectivity.

Here is the dream of a 6-year-old boy, Axel, who is at grips with his Oedipal vicissitudes in relation to his mother who is unable to give him the "man's place" that belongs to him. His parents are separated and he rarely sees his father who shows little interest in him. His problematic place within the family is immediately identified in the first session.

"I dreamt that an iguanodon dinasaur was pursuing me with the aim of devouring me. And yet they're nice in reality. He was breathing fire (iguanodons know, above all, how to swim). All the righters of wrongs were there, Batman, Robin Hood, Superman, etc . . . but they didn't help me get rid of the dinosaur. I slipped on a special suit that enabled me to fly. God gave me a lightening-thrower to fight the animal. The devil suddenly appeared and started to attack me. Then I woke up."

We know that there is a metonymical relationship between water and fire. We also know that the metaphor of throwing fire (or water) is related to the phallic assumption of the little boy. The ambiguity of the dinosaur in Axel's dream refers to the devouring (castrating) avidity of his mother, as is made clear by the child's words: ("And yet they are nice in reality."). It also shows the wish for phallic aggression that the child is seeking to adventize. The righters of wrongs do not succeed in incarnating the paternal metaphor. But God, the object of faith of his maternal grandfather, can. The niceness or kindness in question is in fact the characteristic of the little boy himself, for whom the slightest flight (phallic) is inwardly refused. This is what gives shape to the landscape of the fight with the devil at the end of the dream.

The more an animal is rare, *distant* or inexistent, the more it lends itself to the ambiguity inherent to the *oneiric drive*, and the more it represents for the child the persons who are *closest* to him, namely, his own parents. This uncanny familiarity (*Unheimliche*) would explain the enormous success that dinosaurs have with children who are inclined to project their fears and anxieties.

The playful dimension of dreaming

Are there grounds for ascribing a specific function or precise aim to dreaming? This question cannot be answered immediately. A certain number of premises need to be explored, I think, before we will be in a position to circumscribe our concerns with regard to an eventual teleological aim of dreaming.

We have already discussed a remarkable phenomenon which I described as *oneiric automatism.* This rests on the simple observation that an automatic activity that is often related to images is triggered as soon as we close our eyes. Whether it is reflective or narrative in nature, the activity in question seems to retain its automatic character.

There exist three key moments in sleep: falling asleep, deep sleep (NREM) and paradoxical sleep (REM). The hypnagogical images that appear when we are going to sleep probably do not differ, in their nature, from those occurring during other phases of sleep. They can nevertheless be distinguished from them in terms of their brevity and their associative rapidity. Hypnagogical images occur while we are still in a state that does not allow us to yield to the force of sleep that is trying to overcome us. When we are in the sitting position, for example, on the point of falling asleep, the postural atonia that is suddenly triggered by the brainstem makes us start and brings us back to ourselves, just when we thought we were fighting against sleep. In such a state hypnagogical images are accompanied by postural (muscular) atonia, which is progressive at this preliminary stage. The examples below attest to the common nature of these images with those that occur during dream activity proper (REM).

"I was tired and the book I was reading did not seem very interesting to me. I nodded off and the book fell out of my hands, waking me up in the process. Meanwhile, a precise image had crossed my mind. I was about to throw my book into the *pond* (*mare*) in my garden." We can see that the image in question, just like in a dream, was the depiction of a linguistic expression, *en avoir marre* (to be fed up).

"I was busy writing about brain asymmetry. My eyelids were dropping with fatigue and I saw the image of a naked man who had a sort of swelling on each of his hips. Once I had returned to my senses, I understood that the naked man in the hypnagogical image that I had just noticed was an attempt to represent what

I was reading, that is, the anatomical schema of the human body, the two swellings representing the two hemispheres." Think, too, of the hynagogical image of Mathieu whom I spoke about earlier.[6] The "effigy" of his father, as he was going to sleep, had a high psychic significance, as in his dreams.

It must be admitted, however, that there are not many examples of the kind that I am citing in connection with hypnagogical images. It is often very difficult to interpret these images. In fact, in the phase of falling asleep, many images rush to invest a form of sleep that is not yet fully functional. Sleep seems to be buffeted by the dream imagery that is characterized both by its large quantity and by its brevity. It is as if the dream thoughts were not yet quite in place, for lack of a suitable and well-structured narrative web. Such a narrative organization and structuring come into play, as we have seen, when dreaming fulfils its primary function, that is, the attempt to reduce the excitation due to the partial drives (law of inertia). It is then that the *reverse of the drive* will deploy its supremacy and that the adventization of desire will attempt to assume its definitive form. Remember that the dream process as the reverse of drive activity tries to bring together the partial drives which, owing to their nature, are as fragmented as they are dispersed. The dream process aims precisely at reducing the force of their excitation by bringing them together.

Play between dreaming and waking

If postural control is asleep during deep sleep (NREM), it is paralyzed during paradoxical sleep (REM). In any event, the dream imagery seems to be closely correlated with the motor system. While being dependent on postural atonia, the imagery in question is always tempted to run counter to it in order to invest muscular control. The more the imagery tends towards action, the more the dream work tries to inhibit it in order to preserve the sleeping state. This struggle is sometimes evident in our dreams. We make an attempt to flee, but our legs do not want to follow us. This phenomenon is bound up with the psychic ambiguity with which our dreams are charged and which is revealed through interpretation. This paradox ought to exist at the level of REM and falling asleep but also at the level of slow-wave sleep (NREM). Only it is at its peak in the REM phase. That is why the latter has acquired the name of paradoxical sleep. And it is also why it is so close to the waking state. But it goes without saying that the true paradox is to be found within the law of inertia of the drives.

If dream automatism is at once dependent on postural atonia and runs counter to it, it is obvious that a kind of "play" exists between dreaming and waking. In the domain of sleep, the desire is first and foremost the desire to sleep. We never lose sight during sleep of the fact that we are sleeping. The relation between the desire to sleep and the fact of knowing one is asleep, on the one hand, and that between oneiric automatism and postural atonia, on the other, are both bound up with the principle that the fact of dreaming is related to play. It is very likely that dreams possess a playful logic that we will try to identify from the characteristics that we recognize in them. This playfulness is in conformity with that of the drive. As Freud notes, play and what is serious are not opposed; on the contrary, for the subject the play in question is the high place of seriousness. The subject of the drives "plays", so to speak, or gambles with himself (*se joue de lui-même*); hence his mise en abyme in dreams. This game exists thanks to subjective division. The term division expresses it clearly: there is a gap between the subject and himself, which constitutes the first condition of playful activity.

The playful nature of oneiric mechanisms

The adventization at work in dreaming is in itself a game. Its ludic dimension consists in bringing back to the present, from the past, what is not yet. If we look more closely, it is this same mechanism that governs a child's play. Some authors think that the child is seeking mastery in his play or that he is preparing himself for life to come. These conclusions are certainly not unfounded, but they do not seem to get to the essential issue.

As for dream narrativity, it goes without saying that it, too, is intrinsically related to the ludic dimension. As a primordial element of dreaming, intrigue also participates in oneiric play. It constitutes its desiring dimension. It is the intrigue that keeps the subject of the drives in suspense and that leads him to take himself seriously. Without intrigue, oneiric life would be a state of monotony that the dreamer suffers passively. Without this fanciful entanglement, desire would be a state of melancholy.

Playing is the privileged locus of the division of the subject, promoted during dreaming in the form of the mise en abyme. The latter highlights, as we know, the close relationship of division between the dreamer and the dreaming subject. The mise en abyme suggests the distance that the subject maintains with himself. It creates a transitional space that does not founder in the supposed dichotomy between the subjective and the objective. "*Playing*," Winnicott writes,

> *has a place* and a time. It is not *inside* . . . nor is it *outside*, that is to say, it is not part of the repudiated world, the not-me, that which the individual has decided to recognize (with whatever difficulty and even pain) as truly external, which is outside magical control.
>
> (1971, p. 41, Winnicott's emphasis)

If it is true that playing is an act, it follows that the dream is also an act. In this respect it differs from reflective thinking, and even from the simple fact of desiring. Only in the act, Lacan used to say, is man at one with his signifier. "To control what is outside," Winnicott continues, "one has to *do* things, not simply to think or to wish, and *doing things takes time*. Playing is doing" (ibid, Winnicott's emphasis). We know that for Winnicott, dreaming has the same psychic function as playing. "These two dreams," he writes, "are given to show how material that had formerly been locked in the fixity of fantasying was now becoming released for both dreaming and living, two phenomena that are in many respects the same" (1971, p 31). Following on from this fixity that Winnicott speaks of, it may be remarked that the recurrent dreams of traumatic scenes are as many attempts to transform – to adventize – the trauma into play. This will no doubt allow us to envisage a response to the question of knowing if there is a specific teleological dimension to dreaming. How can one answer this question without forgetting that the essence of playing is its gratuity?

The missing nature of the object

The attempted transformation of the trauma into play seems to comprise an aim to be followed. And yet, as soon as the process of adventizing the trauma begins, we are in a ludic space with the gratuity, but also the seriousness that we recognize in it. We can then witness the see-saw between the antognistic forces. The success of the transformation will depend on the success of the symbolic dimension that is in play, a game, so to speak, that is not ego-based, and where it is not a matter of winning. The aim would be, rather, to unite the conditions which make it possible for the trauma to be integrated within the space of dreaming.

The notion of oneiric automatism expresses it well. It is a form of playing that begins as soon as the eyelids are closed to the outer world. It is nonetheless dependent on sleep which is triggered by the phase of falling asleep. Could dreaming, as an attempt to reduce the drive to the zero level of excitation, be considered as pursuing an aim? Certainly, it can, insofar as it is an attempt at adventization. But it is by virtue of the same tendency that the drive and the reverse of the drive exert their paradox. If there is an aim, it is of a particular kind which does not contradict the playful character of dream activity. Lacan (1973a) refers to it by a word that French borrows from English, the *goal*:

> Here we can clear up the mystery of the *zielgehemmt*, of that form that the drive may assume, in attaining its satisfaction without attaining its aim (*but*) . . . the *goal* is not the *but* either, it is not the bird you shoot, it is having scored a hit and thereby attained your *but*.
>
> (p. 179, my emphasis)

Adventization occurs as an attempt to form a coherent synthesis of the drives with a view to remedying their excitation. Such a synthesis is not an aim in the teleological sense of the term; it is a "goal to be scored" through the *spontaneous* convergence of the partial drives at a suitable moment. This moment may or may not occur. Adventization is always only an attempt, a game.

Several authors have, quite rightly, attributed play with a gratuitous character. Without this character, play loses its ludic essence. Someone who does not respect this principle is called a "bad sport". The principle of gratuity (i.e. something free and done for its own sake) is contested by those who ascribe play with a utilitarian function of adaptation or learning. As for Huizinga, whose book *Homo Ludens* (1938) is a classic in terms of reflection on play, he reduces playful activity to its aspect of competition. *Les jeux et les hommes* by Roger Caillois (1992), another classic, divides games into four different categories: *agôn* (competition), *alea* (chance, hasard), *mimicry,* and finally *ilinx* (dizziness). Caillois divides them further according to their gradual value of *ludus* (game with rules) and *païdia* (spontaneous).

These different categories are not opposed, it seems, to the principle of the gratuity of play, which is the primordial aspect of it. They can be considered as secondary, and particularly with regard to the essential aspects of play which are the gap and subjective division. Play can, in effect, take place secondarily for competitive or adaptive purposes. It then integrates within it a dimension of mastery without, however, losing what constitutes it as such. The dreams involving games of mastery that we meet with in children up until their preadolescence attest to this. But the excess of mastery also risks reaching a devastating drive dimension where the sole aim will result in pure loss. The aim of play will then be relegated to the death drive where *jouissance*[7] will finally exhaust itself in an ever greater repetition.

Far from being a devastating drive impulse, the loss of the object constitutes, on the contrary, one of the constant traits of dreams. The subject is the loser of the oneiric game to the precise extent that his subjectivity depends on this loss. The missing nature of the object is in fact only the token of my division. The failure in question is no other than what escapes me in the Other of myself. The alterity of the subject has a play partner who never ceases to "gamble" with him (*n'a de cesse de jouer de lui*). Play is henceforth the stuff of desire.

In his pioneering ethology, inspired by Husserlian phenomenology, Buytendijk (1920) had already emphasized two principal characteristics of play in living beings. We can recognize here two essential aspects of the drive that has much in common with play. He underlines first of all the ambivalent relationship of animals, that is to say, their movement of withdrawal and attraction towards the object of their play activity; this is what I have referred to as the object's *circumvention of the drive* (see above Part Two, Chapter Two) according to which desire consists of a commencement (*dé-buter*) that is constantly renewed because it is constantly disallowed (*débouté*), which catches hold of us and never ceases to catch hold of us again. What is marvellous about the object of the drive is that it is always missing from its place.

Next he noted youth as a primordial element of play. He highlighted the singular fact that animals that are devoid of a long period of childhood also lack the capacity for play. I mean those animals that acquire almost all their aptitudes at birth or shortly after. In man, childhood continues throughout life thanks to his unconscious drive impulses which make use of dreams and dream imagery. Regarding the human infant, it is against a background of intense playful narrativity, mixing repetition and an ever-renewed desire, that he tries to find answers to the enigmas of his existence. Infantile phantasies, which Freud discovered very early on in his clinical practice, are as many early theories pertaining to the mysteries of life. Evidence of this can be found in the infantile theories on sexuality which, in spite of all the real information put at his disposal, persist surreptitiously throughout the life of the child even as an adult. These theories and phantasies are inhabited by the partial drives which lurk not only in the sexual life of the individual but also in his most remarkable relations with his fellow human beings.

Role of memories in oneiric activity

Freud describes the drive impulses as retrogressive by linking them to infantile memories. "On this view," he writes, "a dream might be described as *a substitute for an infantile scene modified by being transferred onto a recent experience*" (1900, p. 546, Freud's emphasis). We are certainly marked forever by the events we experienced in our childhood, the memory of which has constituted itself as our *having-been*, that is to say, from a point of view that is always present. But the majority of past memories that are reactualized in dreams often possess a paradigmatic character (of loss, reunion, mourning, triumph, castration, and so on), and it is in this respect that they are recalled in dream activity. I will now illustrate this principle with a certain number of dreams.

Juliette is a teacher in her forties. She obtained her swimming certificate at primary school along with her school friends of that time. It had required a lot of effort for her to overcome her terrible phobia of going to the swimming pool which, in addition, made her feel ashamed in front of her school friends. Each time she had to go through a tough ordeal in life, her swimming test reappeared again in her dreams. This was the case during a legal dispute between her and her neighbours concerning a right of way in the property inherited from her parents. Now for another example.

"I was not young. I was going to the hospital to get help for a child. I was looking for the pharmacy in the department of opthalmology. Finally, I found it. I was given a bulky packet which did not correspond to anything for treating eyes." This dream was followed by another. "It involves a child again, but I can't recall any precise details, a child I didn't know well but whom I was looking after. I don't know, but it went on and on forever. The place where we were was dilapidated. I was walking with the child in the corridors. I don't know if it was my child or someone else's."

It is curious that the second dream loses its transparency compared with the first. The associations confirm this. The scene in question is still a hospital, the very one where the dreamer's mother had once had a consultation for her daughter, aged six, who had suddenly become cross-eyed. The medical tests had found nothing abnormal. The doctor had concluded that the condition must have been caused by an emotional shock. The child's nanny had, in effect, just retired and returned to her natal region.

It was the recollection of becoming cross-eyed, linked to the sadness of her nanny's departure, which is at the basis of the dream in a paradigmatic way. The child in question in the dream is that of the first marriage of her son-in-law. The dreamer detests him and finds his company unbearable. He is apparently a chauvinist who takes others as his servants. It is with such scorn that he entrusts his child to the parents of his wife. Our dreamer seems divided between the hatred that she feels towards the father and the tenderness that she shows towards the child, "this bulky packet that is put into her hands". The dream expresses her wish to leave, as her nanny had done in the past, and to leave the child in question in the dream. "Too bad for the child who is in danger of becoming cross-eyed", the dream seems to say. After all, that will be her vengeance on the father. The obscurity of the second dream is simply a result of the embarrassment caused by such hostile feelings. The dream was doubtless an attempt to adventize such gloomy feelings of resentment.

"I was with my father. We were walking along a straight and interminable road. I said something to him but I can't remember what it was. He was shocked, had a malaise, and ended up with a brain haemorrhage. I found myself, with my mother, in front of this inert body. She said: 'Should we call a doctor?'"

The dream inverses what happened in reality. The dreamer had lost, within the space of a year, his mother and then his father. "The straight and interminable road" is the same one he was walking along with his mother when his father had been hospitalized, at a time when he was only five. On the day he learnt about his father's death, he thought to himself that henceforth he was the only one of his family who was still living. This extreme solitude had filled him with a strong sense of nostalgia for his mother. The analysis had clearly revealed his retrospective wish to see his father die before his mother. The fact that he had curiously forgotten what he said to his father in the dream echoed such a wish, finally admitted, concerning his father. The dream took up a paradigmatic memory of an Oedipal relationship dating back to his childhood at the very moment when his own son was preparing to leave him to settle in Canada.

There is quite a widespread opinion that dreams are a regressive process towards childhood memories. But a closer examination shows that there is no solid clinical evidence for such a claim. It should be added that the return of the repressed must be distinguished from the appearance of memories of the past. The return of the repressed always concerns a meaning, a significant link, or a signifier of associative value. A memory does not, in itself, undergo repression. Only the meaning that it conveys can.

It is rare for such an early memory to reappear as such in dream activity. The dream occupies, as we know, a special place in the analytic treatment. Its formation, the fact that the analysand remembers it, and, finally its narrative addressed to the psychoanalyst all play a part in the work of the transference.[8] It is precisely in this respect that the dream promotes the appearance of memories in the deferred action[9] of the situation that is the analytic session. The appearance is due to the lifting of repression, and not the contrary.

Another argument in favour of my thesis – according to which childhood memories do not reappear as such in dreams – is to be found in the Freudian discovery of screen memories.[10] The mechanism of deferred action cannot, in the majority of cases, be dissociated from the process of memory. The memory that we have continually undergoes modifications depending on what proves to be actual in our subjective life.

The major part of the momories that we discover in clinical practice date back to the time after the Oedipal period. Thus the memories dating from the period of latency have often a prominent place in analysis.

A careful reading of the section devoted to "Infantile material as a source of dreams" in *The Interpretation of Dreams* might confirm the foregoing (1900, p. 189). As far as the elements relating to earliest childhood are concerned, Freud does not employ the term "memory" but rather that of "childhood impressions" (*Eindrücke der Kindheit*). Impression should be understood primarily as imprint, mark, or effect. Thus it would be more in keeping with reality to speak of remembering (*souvenance*) and not of memory or recollection (*souvenir*) with regard to the dream elements relating to earliest childhood, for it is often a matter of an ambience, a "zone", and a general atmosphere that give the dream a certain tone and a certain colouring derived from our childhood. These elements prove to be constant in most of the dreams that each of us has.

I would like to add a word on recurring "memories": they are often the convergence of several elements experienced in a condensed form, a new variant of which can be expected each time. They pertain in the last analysis to a drive fixation forming part of a coherent ensemble that constitutes itself as the psychic structure of the subject, whether it is hysteria, obsessional neurosis, psychosis or perversion that is at issue. Even if it dates back to a certain period in childhood, the fixation in question is not for all that a regressive process in the disorganizing sense of the term. It is integrated with the structural ensemble of the subject and constitutes one of its main components.

Circumstantial logic

The so-called regression concerns just as much the constitution of the dream that is the dream work.[11] It is the latter, in particular, that gives it its bizarre and incongruous character. The result is sometimes a particular system of reasoning similar to what Allan Hobson (1988) calls the *ad hoc* logic of the dream (p. 266). This logic confirms one of the conditions, not regressive but playful, of the dream. It consists in establishing links and lines of reasoning which prove to be childish and without foundation once they are removed from their context. It has the appearance of a momentary convention so that everything holds together, even if for just an instant provided it is long enough, so as to be able to accomplish the oneiric act.

The circumstantial logic of the dream consists of a set of "game rules" that could be modified if it was deemed necessary. The two following examples may help to elucidate these remarks further.

Marine is a young girl of thirteen. Owing to her parents' difficulties, she had been placed in a foster family. Suffering from the separation from her parents, she was constantly accusing the foster family of being at the origin of her misfortunes. During a session of analysis, she reported the following dream: "I was in a park. I saw a lady passing, accompanied by her dog. It looked nasty. I tried to turn my head away so as not to attract its aggressivity. I felt that it was still looking at me nastily. I turned round. Strangely, I saw it immobilized in the frame of a painting that the lady was carrying under her arm." We can see that the girl's terror when faced with the dog was such that dream changed its logic and transformed the dog into an inoffensive image. The analysis was able to determine that the lady was no other than her foster mother, the favourite object of her fiercest projections. The lady's hair in the dream was in effect an allusion to her family name. "Tricks" of this kind are numerous in oneiric activity. Here is another example.

Claudia is a woman who is somewhat advanced in age. She is concerned by her daughter's pregnancy. She doesn't care much for her daughter's companion and the idea of having a grandchild as a result of their union is unbearable. "I dreamt that I was slipping into a piece of pottery which I have had in reality since my daughter's birth. I managed to get it to roll along the ground before getting broken. At the bottom of the slope, I was suddenly afraid. Fortunately, it had come to a standstill without getting damaged." The piece of pottery is of a particular kind and has a Provençal name evoking the idea of pregnancy. The fact of wanting to break the piece of pottery is an aggressive impulse directed at the young couple. We see that at the last moment the aggressive logic changes so as to keep the dreamer's life intact.

We will return later to the question of these "game rules" that allow us to modify the way our dreams unfold. First we are going to study other points concerning dream activity, beginning with the enigmatic relationship between play and chance.

Notes

1 *Cf.* the law of inertia in the Freudian theory of the drives.
2 See Freud, 1900, *The Interpretation of Dreams,* Chapter VII.
3 See Part One, Chapter One.
4 "At the first reflective glance," Sartre writes, "we see that we have so far committed a double error. We thought, without justifying it to ourselves, that the image was in consciousness and that the object of the image was *in* the image. We depicted consciousness as a place peopled with small imitations and these imitations were the images. Without any doubt, the origin of this illusion must be sought in our habit of thinking in space and in terms of space. I will call it: the illusion of imminence." (Sartre, 1940, p. 5, author's emphasis)
5 Term that refers to an intense affective cathexis by the patient of the person of his/her analyst.
6 Part Two, Chapter Two: "Dream work and the neurophysiology of sleep".
7 The term *jouissance* (overflow enjoyment) designates for Lacan the surplus value of instinctual drive satisfaction. In this respect it invades the subject's universe entirely. It is thus bound up with the death drive, which is characterized by repetition.
8 The transference (*Übertragung*) is the "link establishing itself in an automatic and actual fashion between the patient and the analyst, reactualizing the signifiers that conveyed his demands for love in childhood . . ." (Chemama and Vandermersch, 1998).
9 Deferred action (Nachträglichkeit; Fr. après-coup): "Term frequently used by Freud in connection with his view of psychic temporality and causality: experiences, impressions, and memory traces may be revised at a later date to fit in with fresh experiences or with the attainment of a new stage of development. They may in that event not only be endowed with a new meaning but also with psychical effectiveness." (Laplanche and Pontalis, 1967, p. 111)

10 The screen memory: "A childhood memory characterized both by its unusual sharpness and by the apparent insignificance of its context. The analysis of such memories leads back to indelible childhood experiences and to unconscious phantasies. Like the symptom, the screen memory is a formation produced by a compromise between repressed elements and defence." (Laplanche and Pontalis, 1967, pp. 410–411)

11 Freud calls the dream work "the whole of the operations which transform the raw materials of the dream – bodily stimuli, day residues, dream-thoughts – so as to produce the manfest dream" (Laplanche and Pontalis, 1967, p. 125). For Freud, distortion results principally from the dream-work.

9

ONEIRIC REPRESENTATION, DRIVE, AND ITS OBVERSE

Means of representation in dreams

The game has its own rules each time. It marks its limits in accordance with the vicissitudes which drive it. Freud has taught us about these rules. The dream, he says, presents logical relationships as simultaneous. When it brings two different elements together, it establishes a link between them. On the other hand, causal relationships are shown in their succession. Thus the dream prologue and the principal dream often stand in a causal relationship to each other. To give meaning to the "either/or", it makes the two terms of the choice equivalent. The unknown person in the dream, he indicates, is often none other than the dreamer himself. We understand then that the mise en abyme of the subject is at its height.

The representability (*Darstellbarkeit*), more often called *figuration* in French, consists of transforming the signifiers and linguistic expressions into dream imagery. Thus the expression "*faire ailleurs*" (lit. "do it elsewhere") is in a dream transformed into the image of the "*ferrailleur* (scrap merchant)", which sounds the same as "*faire ailleurs*" phonetically. By such an image, the dreamer would invite his young son, who was about to have sex with his girlfriend, "to do it elsewhere". In yet another dream, the young girl says that she is attending a wedding even though the scene is veiled. The interpretation established that it was none other than herself wearing the veil of the bride. We can see then that an image (the manifest content) hides the signifier of the dream (the latent content). The reader will find below a more complete range of the means of representation employed in dreams.

A young man aged 25 reported the following dream, which occurred between two awakenings in the morning. "I was thinking about the covering that I was intending to put on my garden shelter, which I was going to busy myself with during the day. I was wondering whether there was enough to cover it completely. I then slipped again into sleep dreaming about wearing trousers that were too long." The excess length of the trousers was undoubtedly related to the dreamer's concern that he had prior to falling asleep, that is, the length of covering he needed for the shelter which he expected to finish during the day. If that is the meaning of the dream, the question that then arises is to understand why the trousers replaced the covering? The dreamer was no doubt relating, by a curious form of mise en abyme, the construction of the shelter to his own body, while satisfying his desire in an exaggerated way, inverting the "too short" (of the covering) for the "too long" of the trousers. One would then be entitled to ask oneself why the mechanism of the mise en abyme of the subject should prefer the trousers over anything else linked to his body. Certainly the signifier "covering" must have encouraged the passage between the two entities, but it does not seem to fully cover the question we have been putting as to why the choice of trousers caused the mise en abyme of the subject. When the young man was asked about his physical state when asleep, his reply gave us the expected clarification. He explained that when he was asleep, he had had a pressing need to urinate, a problem, he said, that resembled the embarrassment of putting on such long trousers.

The dream highlights the principle according to which the dream depiction or representation (*figuration onirique*), that is to say, the transformation of the signifiers into image form, is often permeated by the mechanism of the mise en abyme of the subject. The covering intended for the garden shelter is only transformed into trousers to the extent that the subject appears in the dream through the medium of his own body. In the same way, the wish is only satisfied here to the extent that it is projected from a future lived in the present time. The planned covering of the shelter is brought back to the present, which moreover is brought on by the urgent need to deal with the bodily needs, albeit delayed by the need to continue to sleep. The troublesome length of the trousers is substituted, then, in the form of a dream image, for the imperative of the bodily function in order to prolong the desire to sleep. The dream is in effect the guardian of sleep.

Freud tackles the question of representability *(Darstellbarkeit)* in Chapter VI of *The Interpretation of Dreams.* He explains there the different means available to the dream in order to express in dream imagery the various linguistic processes:

> It is fair to say that the productions of the dream work [*figuration*], which, it must be remembered, *are not made with the intention of being understood,* present no greater difficulties to their translators than do the ancient hieroglyphic scripts to those who seek to read them.
>
> (1900, p. 41, Freud's emphasis)

We know that *The Interpretation of Dreams* is unquestionably the confirmation of the Lacanian maxim according to which the unconscious is structured like a language. Consequently, pictorial representation (*figuration*) demonstrates how the dream image puts itself in the service of unconscious speech.

This is the dream of a young twin girl aged 17. "I was in an old-fashioned lift. I was squeezed in tightly and had a strong feeling of suffocation. There was a kind woman beside me whose presence comforted me." Then came the associations. The lift reminded her of the apartment block in which one of her old school friends lives; he had been in love with her, she said, since the age of 11. The life this young man was leading now consisted in spending all day in his maid's room in front of his computer screen. He only went out to do his shopping. "His life," she said, "frightens me; it makes me think of the life I was leading before I began to work on myself in therapy". The kindly woman undoubtedly conjured up her mother. She reacted positively to the following interpretation that the lift (*l'ascenseur* in French) is a phonetic expression of "[*la*] *sans soeur*" (lit. "[the] without sister"), saying that "the boy I was talking about was obsessed by the close resemblance between me and my sister. He irritated me with his fixation on us as twins."

The dream reflects the analytical path on which she had embarked with the aim of finding her true self, as distinct from her sister. It seems to say the following to the boy who is only interested in her as a twin: "I am living under my mother's goodwill without being stuck the whole time to my sister. Will you understand that if you want to take interest in me, you must put aside the matter of my twinhood and consider me for what I am?" Now we understand that the dream is trying to adventize her problems with twinhood. We are going to find below other examples of the figurative production of the dream work.

Here are two dreams that occurred during the same night. As Freud advises us to do in similar cases, it is by reversing their order that we come to penetrate their secrets best. The child involved in these dreams is that of the first marriage of the dreamer's son-in-law, whom she considers *persona non grata*.

1. "I dreamt that something was flowing out of me. I was losing something. I looked at myself in the mirror. Blood was flowing from my ears." We understand, after the telling of the second dream, that the first either has a punitive impact in an introjective form, due to the agressiveness expressed, or a precise intent aimed at clouding the issue so that we cannot find the real "culprit".

2. "There were various kinds of lemon-squeezers, small, large, old, new . . . There was also a child with me. Did I need to press something for him? Suddenly, there were lots of people. I looked for the child. He had disappeared. Panic stricken, I woke up." There followed a series of denials that were designed to conceal the violent content of the dream. Despite establishing the connection between the lemon-squeezers and the expression "*presser le citron à quelqu'un*" (squeezing someone dry) the dreamer continued to reject the aggressive nature of the dream. It is necessary to fully recognize that such a "crime" could only be shameful.

Given the aggressive wish in the two dreams which, moreover, is fulfilled, there is reason to ask oneself where their aim of adventization lies. Such dreams seek in fact to adventize a violence which would otherwise be too burdensome for the subject. They want to be brought to the analyst's knowledge, something which in itself brings a degree of relief, a certain distance. This is often the real reason for their dream formation during analysis; otherwise they would have stayed outside the sphere of dreams. Here now is another dream depiction following obsessive ideas during the night.

A single woman has started a relationship with a divorced man who does not hesitate to mistreat her. The following dream occurred during the night when she first had sex with the man in question. She had spent the night virtually sleepless, turning over in her mind obsessive ideas about him. The man's brutality was not limited to the physical, but was also on the moral level. He never stopped abusing her verbally throughout their love-making.

"I put my cat (*chatte*) into the washing machine." This summed up the essence of her dream in the night. The cat was clearly an allusion to her genitalia (also *chatte*, in slang) which were particularly sore; but its washing was an attempt to adventize the moral stain she felt she had incurred through the verbal violence inflicted on her by her partner. What could be more eloquent in this respect than the following association? "In a comic strip, a cat and a dog are washing the floor. They take themselves as mops and dry themselves on the washing line." We recognize then the signifier "mop" (*serpillière*) in all its semantic ambiguity.[1]

"I dreamt that my friend René had learned that he had been struck down by an incurable illness. He was 'celebrating' his own funeral ceremonies. Without there being anything incongruous or absurd in the dream, the 'celebration' was in full swing. Everything carried on as if it were a normal celebration. It took place in a house of one of my long-standing friends. The reception was held in this house, but also in another place, on some sort of island or a plateau, dominated by the sea."

The first thing that catches our attention is the link between the first name of the friend in question, René (lit. "reborn" in French), and death. The fact that he is a doctor overdetermines the link that the dream establishes between life and death. The layout of the house, where the dream takes place, leads us in the same direction. "It was, in reality, a house whose basement opened out on one side, but whose ground floor opened out on the other side, so that one entered the house from the road behind or from the road in front." The basement reminded him immediately of death, but the link, one sees, does not end there. The father of the "long-standing" friend at whose house the "celebration" took place, had for a long time haunted the fears and anxieties of the dreamer. The father had left his country to seek treatment in Paris. He had died there in total exile, knowing absolutely no one, neither close nor far. This memory must have been brought back to him by a visit to the Père Lachaise cemetery accompanied by one of his friends on the eve of the dream. "I had accompanied him to the cemetery in question. He wanted to pay hommage to a writer from my country of origin who had committed suicide in Paris."

The island, or plateau, dominated by the sea, "reminds me both of the country of origin of my friend René, and of a video game belonging to my children, which takes place moreover on a three-dimensional cosmic plateau, which accentuates its sense of vertigo and anxiety." René is, so to speak, the alter ego of the dreamer. His presence in the dream corresponds to the mise en abyme of the subject/dreamer. They are the same age. There is close socio-political affinity between them. The first name of René's wife suggests, too, the idea of birth. The latter is in effect the principal issue of the adventization of the dream, and this was after a bitter period of despair and torment. The dream seems to say, "Do not fear death, profit from the time that is given you." This idea is confirmed by a major association; René is the male version of the name of the former partner of the dreamer. It required a big effort to separate from her. She represented for him the proximity of death. It is hardly difficult for us to grasp the *ad hoc* logic of the dream: the celebration which replaces the bereavement, the funerary ceremonies which precede death, the house which is at the same time a *cosmic* plateau . . .

The dream is a paradoxical game. It wants to bring everything into the present with a view to adventization, but makes the future disappear. According to the Freudian point of view, the present time is the locus of conscious investment. Thus it is that the dream is close to the state of wakefulness to which it so often leads.[2] But this seeking for the present and consciousness is opposed to the attempt by the dream to preserve the sleep on which it is dependent. The paradox lies precisely in the fact of remaining in its dream state, while wishing to invade the conscious mind. It is trying to adventize either that which is no longer in its full presence or that which has not yet come into being. This movement reaches its peak when we notice that the dream is the gathering together of the past and of the future and that the *adventization* consists precisely of going towards that which is not yet and of that which is no longer.

The oneiric paradox

The idea of the dream as drive has led us to consider oneiric activity as a paradoxical movement which goes against the drive from which it emanates. We know that the law of inertia which governs the drive is closely bound up with the purpose of sleep which is rest. Thus it is the first aim of the dream to reduce the partial drives, at work during sleep, to their lowest level of excitation. It is here that paradoxically an intense cerebral activity occurs, in the form of dreams, in order to calm the flurry that has arisen. This paradox is, we have seen, inherent to the mechanism of adventization. The latter, in fact, pursues its objectives in accordance with the law of inertia put forward by Freud in relation to the drives. Consequently, the drive contains within it its own opposite. So the dream as adventization requires an intense cerebral activity, without which it would not be able to remedy the excitation of the partial drives. The process whereby the adventizing activity of the dream takes charge of these partial drives can take place at any phase of sleep, although that of paradoxical sleep is clearly the most favourable.

Freud's distinction between the dreams "Autodidasker" and "Father, don't you see I'm burning?", one ideational, the other hallucinatory, would correspond to that of the neurophysiologists of sleep between REM and NREM. Freud maintains that "there are dreams which consist solely of thoughts, but which cannot on that account be denied the *essential nature* of dreams" (1900, p. 535, my emphasis). The question that arises is to know why he considers these entities as being dreams even though they lack any hallucinatory quality. This absense of transformation of dream thoughts into images is present, says Freud, "in every dream of considerable length, which are simply thought or known, in the kind of way in which are accustomed to think or know things in waking life" (p. 535). Should we understand that for Freud the essential point of the dream resides in its fulfilment of the wish and its dissension from censorship? But we know that these elements intervene similarly in other formations of the unconscious. They are also constituents of phantasy and daydreams.

As for censorship, Freud does not give it an exclusive role in the formation of dreams. Censorship is perhaps a necessary factor in oneiric activity, but is not, of itself sufficient to characterize the dream. He writes:

> If what enabled the dream-thoughts to achieve this (access to consciousness) were the fact that at night there is a lowering of the resistance which guards the frontier between the unconscious and the preconscious, we should have dreams which were in the nature of ideas and which were without the hallucinatory quality in which we are at the moment interested.
>
> (1900, p. 542)

While noting that the hallucinatory aspect is not exclusive to the dream, Freud noticed that wherever it is present, it occupies the foreground. He cites Fechner who "*puts forward the idea that the scene of action of dreams is different from that of waking ideational life*"; and Freud adds that "this is the only hypothesis that makes the special peculiarities of dream-life intelligible" (p. 536, Freud's emphasis).

The question which arises is to understand why, after having demonstrated that dream imagery stems essentially from language and that the principal objective in dream activity is to put unconscious speech into image form, Freud comes to question its hallucinatory character. Why does he come to a stop in such a late chapter in front of the sensorial image of the dream? Because the image is a constituent part of the dream. When Kleitman and Aserinsky highlighted the correlation between REM and dreams, they had noted a major characteristic of dream imagery which remains to this day a subject of investigation.

In dreams all our links to the outside world are conserved. This is above all because we are capable of seeing and looking. But what sort of vision and what sort of looking is involved in a dream? Once our eyelids are closed, what is the range they give us access to? Freud speaks of a psychic locality, of another scene. He places the latter at the ideal point of an apparatus which is totally optical. "Psychic locality (*psychische Lokalität*) will correspond," says Freud, "to a point inside this apparatus at which one of the preliminary stages of the image comes into being. In the microscope and the telescope, as we know, these occur in part at ideal points, regions in which no tangible component of the apparatus is situated" (1900, p. 536). It remains, however, that he excludes any "anatomical localization" for this psychic locality, in order to build his outline of a psychic apparatus.

The intensity of oneiric images

With regard to the psychic apparatus whose outline is discussed in the last chapter of *The Interpretation of Dreams*, McCarley (1998) accuses Freud of iso-morphism. "Both Freud and other theorists of his day, believed in direct, simple correspondences between brain and mental events, and thus it is not astonishing to find the *Project* and *The Interpretation of Dreams* replete with these corre-spondences, which we refer to as isomorphisms" (p. 116). For sure, as McCarthy rightly points out, the outline of the psychic apparatus in Chapter VII comes from the *Project,* which was intended to build a psychology based on neurol-ogy. But he disregards the fact that Freud's intention is totally different in *The Interpretation of Dreams,* and that he is even, in some respects, close to the pre-occupations of those who uphold the *activation-synthesis* theory.

In fact, Freud's examination in *The Interpretation of Dreams* concerns the imagery of dreams. His outline (see Figure 9.1) portrays the psychic apparatus in a succession of subsystems beginning with perception (P) and ending with motil-ity (M), via the mnemic traces (S^1, S^2 . . .). The usual direction of the apparatus goes from perception (P), conceived under the mode of a reflex system, to motil-ity (M) as the final discharge of energy. But according to Freud the hallucinatory dream follows, in the above apparatus, a *backward* direction, that is to say in the opposite direction to the arrows. "Instead of being transmitted towards the motor end of the apparatus, it moves towards the *sensory* end and finally reaches the perceptual system" (1900, p. 542). Why does Freud feel the need to put forward such a set-up and why does he resort to the notion of regression? It is here that we discover what it is that is troubling him. It is in order to explain what he calls the full sensory *vividness* of dreams. His explanation scarcely hide's the questions that are troubling him. "We have done no more than give a name [regression] to an inexplicable phenomenon" (ibid., p. 543).

The Freudian hypothesis of the retrogressive path of dream images could be considered as being in line with research into the neurophysiology of sleep. The correlation discovered between the REM and dream images by Kleitman and Aserinsky in fact reinforces Freud's explanation with respect to the vividness of hallucinatory dreams. Closer to us, neurophysiological studies seem to have

FIGURE 9.1   The psychical apparatus according to Freud (*The Interpretation of Dreams*)

suggested a certain functional gap between the visual stimuli coming from the retina towards the thalamus and those arriving at the latter from the visual cortex. This gap is very close to that which we notice in the phenomenon called binocular rivalry in dissimilar images. The retro-connections of the nerve cells between the visual nerve-centre and the thalamus seem then to have a greater amplitude than the connections between the retina and the cerebral relay that is the thalamus. From this we understand easily that dream images arising in the absence of stimuli from the outside world are of greater intensity during sleep.[3] Also, the research of the American neurophysiologist, Rodolpho Llinás, which we discussed before,[4] seems to weigh in favour of the Freudian hypothesis.

The concept of regression once again highlights the fact that Freud did not regard dream censorship alone as the origin of the bizarreness of dreams, despite the criticism of McCarley who rejects the Freudian theory of the disguise of repressed wishes by turning his back on Freud's questions concerning the dream image. "If we regard the process of dreaming," Freud says, "as a regression occurring in our hypothetical mental apparatus, we at once arrive at the explanation of the empirically established fact that all the logical relations belonging to the dream-thoughts disappear during the dream-activity or can only find expression with difficulty" (1900, p. 543). To put it another way, the distortion of dream events which occurs on waking up, as well as our tendency to forget them, would, according to Freud, be attributable to this retrogressive path during sleep which he calls regression. This is in fact an argument of the greatest importance. Consequently, it would be worthy of interest to see if later neurophysiological research manages to prove this.

Let us return to the question of the intensity of dream images. The dream can also use the intensity of these images as a specific way of expressing unconscious desire. Hence the intensity of the image will also play its part in the rules of the oneiric activity.

"I dreamt of a tomato. It had a bright red colour. It was not for eating. It was there simply for its beauty."

This was the dream of a woman, barely past her fortieth birthday, whose analysis had succeeded in ending her alcoholism. After her detoxification, she gradually came to understand that her real addiction was nothing other than the person of her mother. Aged three or four years more than her sister, she had retained from her childhood the traumatic memory of having seen her mother reappear, with a baby at her breast, after having suddenly disappeared for what had seemed like an eternity.

Her mother was a very domineering woman, who imposed her supremacy wherever she could. She quite literally had her daughter under her thumb, despite her apparent disobedience. Her realization of the degree of control her mother exerted over her led our dreamer towards the acting out that signalled the breakdown of the relationship. The dream above happened at a moment of intense nostalgia for her mother. This feeling ended up in exacerbating her need to drink. But the weaning was nevertheless respected. The tomato, in the dream, is in fact, a metonym for alcohol. In a period of total alcoholism, a doctor had suggested to her "that a glass of fresh tomato juice would certainly help the longing for alcohol to pass". Thus the tomato in the dream was deprived of its normal function; it was not there to be eaten, but just to be looked at. Its beauty and its bright redness said a lot about the oral impulse which had not lost its intensity since her childhood. That was also an allusion to the irrepressible wish to find again the beauty and the presence of a mother who had so effectively nourished the love that was consuming her daughter.

As we can see, the vividness of the oneiric image in this dream forms, in itself, a dream process stemming from the desire which underpins it. We will see the same phenomenon at work in the dream which follows. This concerns a young girl aged 16 who has fallen deeply in love with her friend's lover.

"The setting is springtime; a sort of lake had been formed by rainwater which had no doubt just fallen. I was swimming with ease in the middle of the lake and was really enjoying moving about in the water. I got out at a specific moment and found myself in front of a handsome boy with deep black eyes. He looked at me tenderly for a long moment. I am struck by the colour of his eyes. I soon realized that he was the fiancé of my sister, who suddenly appeared at my side."

The account of the dream was followed by an association which seems to have been decisive in the formation of the dream. One day she had heard her friend's lover use the expression "*se noyer dans le regard de quelqu'un*" (melting [here, lit. "drowning"] under someone's gaze). The aquatic metaphor seems in fact to be the colourful expression of the scene which ensues. She literally melts under the gaze of the handsome boy, which was, however, intended for her sister, a substitute here for her friend. Thereafter, despite the immense tenderness she feels for him, she resigns herself to leaving him to her sister, who is for her a paradigm of all feminine rivalry. Hence the sense of well-being she feels in the water as an expression of the relief she feels after the grief caused by her shameful love.

Oneiric presentification and the vividness of the image

Let us return now to the question of the vividness of the dream image. Should we not rather look for the answer in the specific nature of the dream temporality as it determines the occurrence of the images? These in fact turn out to be invested with a quite specific intensity turning towards a present time which evokes the overflowing of the drives in accordance with their vicissitudes. This exacerbated time is part of the state proper to paradoxical sleep. And it is without doubt in such an intensity of presentification that we may search for an explanation of the vividness of dream images. The storm thus produced by the drive movements is such that everything is brought into the present, in defiance of all conceptions of linear time. It is here that the mechanism of adventization can only remain at the level of an *attempt* aiming to reunite the partial drives, thus activated, in order to give them intention and purpose. What we see, in fact, during paradoxical sleep, is the boiling over of the drives bringing with them a confused mixture of unconscious thoughts and memories coming from everywhere in order to take control of the special means that is the specific state of paradoxical sleep. Then everything is in a hurry to enter into conflict with everything else. Not only the dream scene which has arisen is well and truly in contradiction with sleep, whose maintenance is indispensible to it, but also the wishful thoughts, brought back to the present time following the agitation of the drives, do not always prove to be, indeed are far from being, in accord with each other. They are even at risk of becoming tangled up, compromising the pursuit of sleep. The moment when the dream occurs is one that is both full and intense. It is an immense leap into the past, not to repeat it, but to *liberate* it from the forces contained in it. *The dream is an awakening with a start from the past*, a past as a result that is reinvigorated and barely recognizable thanks to the brightness of its *novelty*. But its success will depend on the degree of working-through taking place in the psychic progression of the subject at the moment when the dream process unfolds. It is, moreover, for this reason that the dream is always a precious indicator in our clinical work.

The images that occur during sleep are not simple reproductions from the past. The intensity deployed by the archi-presentification specific to the dream grants them a new configuration which until then had been in a latent state. On the strength of recent upheavals, they are now in a position to manifest themselves in a manner that is especially intense and in a new light. A new horizon then emerges which enables them to become part of another purpose, and all this is thanks to the process of adventization which is set in motion.

Similarly, the dream image, as vivid as it may be during sleep, is never the simple reproduction of a reality that existed in the past. Its occurrence in the dream raises it to the rank of *figure*. It is, moreover, in this capacity that it will be able to transform itself into a dream *figuration* (*Darstellung*). It is in procuring a new horizon for the latter that adventization manages to raise it to the rank of a *configuration* that is the dream itself.

The dream figure is a present time condensed from simultaneous images. It oscillates continuously beween the visual entity which it is and the signifier that it is able to represent. It cannot simply be reduced to the status of an image. The signifier dwells in the dream figure in a silent form. Charged with a drive impulse that is above all devoted to the visual, it cannot be assimilated with the image as a carbon copy of external reality. It is by allowing itself to be described by the subject that it reveals totally naturally its linguistic dimension.

The obverse of the drive

The present moment is the right time for the partial drives. Hence their increased intensity during sleep in the course of which hallucinatory dream images are most vivid. Thus it was that Freud assigned to the dream a specific place that he described as the other scene *(andere Schauplatz)*. This is not only a privileged site for the drives to manifest themselves, it itself constitutes a drive force governed by the oneiric automatism whose first characteristic is related to hallucination, that is to say, to the dream vision. This vision embodies what Freud calls, in the case of the drive, the joy of looking (*Schaulust*). Once the eyelids are closed, this impulse to look or to see acquires a character that is even more intense. The partial drives will then try to take over this favourite place in order to give free range to their expression. Freud himself speaks of the "concomitant attraction" or of the "*selective attraction* exercised by the visually recollected scenes touched upon by the dream-thoughts" (1900, p. 548, Freud's emphasis).

The dream is the paradigm of the desiring vision. As a drive, it is a hallucinatory process. This means that oneiric impulse is essentially figurative. It is here that the transformation of thoughts into visual images (*figuration*) in the sense of theatrical representation (*Darstellung*) stands in for presentation or idea (*représentation*) in the cognitive sense of the term (*Vorstellung*). Such visual representation is itself a source of exacerbation for the partial drives. It brings them to their highest state of excitation. Hence the adventization which will try to calm, in a more or less successful synthesis, their state of agitation that has reached saturation. The attempt in question, says Freud, will function as a purposive idea (*Zielvorstellung*)[5] while making use of the vicissitudes[6] of the partial drives. Thus the dream brings these to their peak. It is here where dream activity, as a paradoxical entity – at one and the same time true to its active drive dimension and in conformity with the law of inertia – will start to correct this excess of excitement. It is thanks to its tendency to reduce drive excitation to its lowest level (law of inertia) that the oneiric drive will bring its own logic into play. This includes the totality of the dream devices which we have discussed before, i.e. narration, intrigue, the mise en abyme of the subject, and so on. The mechanism of adventization also uses the vicissitudes characteristic of the partial drives to act as their purposive idea. This last goes beyond the laws controlling the phenomenon of association, in the asssociationist sense of the term, in order to give the partial drives an aim (*Ziel*). The purpose in question helps all the partial drives that are present to converge towards the sought-after satisfaction. But the satisfaction aimed at is nothing other than adventization as a remedy for the excitation that occurred in sleep. It is then the obverse of the drive in accordance with the law of inertia. It is essentially this tension at the heart of dream activity between the drive and its obverse which engenders the constant ambiguity that we notice in dreams.

It is the process of *visual representation* (*Darstellung*), that allows the dream process to take charge of the drives. But this is only possible thanks to the aptitude of the partial drives themselves to be treated in this way by their vicissitudes. This is what Freud calls attraction: "the transformation of thoughts into visual images [*figuration*]," he writes, "may be in part the result of the attraction which memories couched in visual form and eager for revival bring to bear upon thoughts cut of from consciousness and struggling to find expression" (1900, p. 546).

The oneiric drive is the convergence of the partial drives revealing themselves, higgledy-piggledy, with a view to adopting the pictorial images made accessible during sleep. Yet, although it conforms to the essence of the drives, this process constitutes a first condition for repressing the partial drives. It is this which grants them access to the dream logic and acts as the necessary condition to enable them to be integrated into the narrative framework of the dream with a view to adventizing the unconscious wish. In this way, a general movement is created during sleep leading these scattered drive impulses towards a synthesis whose aim is none other than to act as purposive ideas.

The movement of adventization is already under way at the time when the partial drives of which it is composed reveal themselves. It is their convergence that forms the fabric of dream logic which gives rise to the dream narrativity. Adventization is built on the principal aim of the oneiric drive which tries to calm, that is to say, to reduce the storm of drive excitement to its lowest level. It is then that the drive is transformed into its obverse. Without the intervention of such a motion, the dream would risk causing a total upheaval threatening the continuation of the sleep on which it depends.

A close examination of dream activity will lead us to understand that the adventizing purpose is already under way when the partial drives are represented in the dream. This boils down to saying that the convergence of these drives as purposive ideas is not a motion which is added from the outside or in the after-math of the adventization. It is not a total drive set in opposition to the partial character of the drive impulses which precede it. There is no global drive that totalizes the experiences of the individual. However, the essential nature of man demands that he aspires to such an ideal. It is just such an aspiration which determines his temporality as waiting-for and his forward march as adventization.

The object of the oneiric drive

One of the most important elements in the drive is its object. How does this manifest itself in dreams? Under what form could it appear following the principle of dream representability? Answer: in no specific form, but in innumerable ways. It is not a case of such and such an object which could be cathected by the drive impulses, however prominent its affinity with the latter may seem. The object in question has the particular feature of eluding our gaze. That is why it necessarily escapes every gaze, including our own oneiric gaze which nonetheless does everything possible to reveal it. The representability (*figurabilité*) of the dream turns in fact around such an impossibility. It reveals it in multiple forms without managing for all that to capture it in an image. Such an attempt plays a part in the ambiguity, the obscurity, and the bizarreness of dreams. In such an impossibility, the object of the drive, around which the entire oneiric drive crystallizes, presents the enigma of the desire of the Other.

Freud describes what is in question as a lost object, an object which has never existed even if it is a matter of refinding it. As the cause of my desire, it divides me by making me the Other of myself. It is this of which we have no idea (as in Lacan's theory). It is neither specular nor representable. It is the eternal missed encounter.

The object, says Freud, is the part that is detachable from the body. Let us take as an example the oral impulse: it harbours within it ambivalence par excellence. It consumes its object with love while wishing to conserve it. Hence its real object is not the maternal breast. If there is an oral object, it is somewhere between the envy and the gratitude of the baby (Klein, 1957), between the love and hatred where the subject of the drive, divided, finds itself, embracing the "nothingness" of its object.

Dream depiction is the constant attempt to represent an object of this kind. When this attempt fails, the partial drives start to converge with each other in order to be able, with the help of the dream logic, to express it as an ellipse, that is to say, that which can only be read between the lines. Thus it is that dreams always require interpretation. These vicissitudes would not exist if there were a genital drive that totalized the partial drives as a whole. Above all, they would not exist if the "*objet petit-âme*"[7] had a visual, representative or specular consistency substance.

This would appear contrary to that which constitutes the essence of the subject as being ahead of himself. Certainly, there is not an overall drive which would embrace all the partial drives in one totality aimed at remedying the fragmented character of the latter. No drive exists which would satisfy the subject by putting an end to his constitutional non-completion. But we must not lose sight of the essence of man which resides in his aspiration for such a gathering together. And it is here that we find the whole raison d'être for the attempt at dream adventization.

In the dream, are we gaze or vision? The mise en abyme of the subject, which is an essential factor in dreams, teaches us that it is a case of looking at oneself-seeing (*se regarder-voir*). This is not an ego-based looking, but a looking that is blended with the act of seeing. In hypnagogical images we witness the "unreeling" of images, for lack of narration, while, in the dream proper, we are all seeing, a seeing which looks at itself. The narrative intrigue of the dream is made up of this close interweaving of the gaze and seeing. If we watch ourselves seeing in the dream, it is because it is a gaze that tends to circumscribe, with a view to adventization, desire that has become vision. It is possible that dream censorship sometimes gives a fearful tone to the fabulous aspect of the dream. In dreams, we are often torn between fascination and fear, two exacerbated features of the gaze.

In the vividness of the dream image, what leads us into error, without even realizing it, is the comparison we make between looking in the dream and looking when awake. The dream image is not the object of the gaze. We are included as vision. But this vision, however, is directed at something. Its aim is no longer object-related. It is rather a letting oneself be led by that which keeps us waiting, perhaps at the risk of an intrigue concerning the advent of the object, cause of our desire. And this in the state of intrinsic ambiguity in which we are faced with ourselves, having become another from whom we are waiting in fear, and sometimes sneakily, for the expected which is the object. Waiting is consubstantial with desire to the extent that the latter is itself grappling with the impossibility of the real as a missed encounter. Thus all desire is intrinsically directed towards adventization. Waiting is rich in events and intrigues which bring it to its peak without attaining its object. It tightens like a bow which places its two ends in tension, the past and the future, towards a point of dissension called the present, an illusion of reunion. The present is set up as the point of tension between the two extremities in order to form the intrigue where, as desire, the dream narration is knotted; the latter being the *mise en scène* in which the director and the actor are the same and the Other.

Notes

1 Translator's note: *serpillière* may also have the sense of servility, a wimp or wet rag.
2 Cf. dreams occurring in the early hours of the morning.
3 See Singer and Varela (1987); and also Singer (1977). Readers may also like to consult the interesting doctoral thesis of Adrien Chopin on at http://adrien.chopin.free.fr/papers/version_avancee_5_09.pdf.
4 See Part Two, Chapter Two, "Dreams and the neurophysiology of sleep".
5 This is where Freud moves away from the traditional idea of representation. According to the philosophical tradition, in particular since Locke and Hume, ideas or presentations (*représentations* in French) are governed by an associative principle. Three essential characteristics govern the latter: contiguity, ressemblance, and causality. But according to Freud's conception, associations are oriented, in their internal sequence, by a *specific purpose* which guarentees their constancy. The sequence in question is not for Freud a mechanical phenomenon but is characterized by certain key markers. The latter draw towards them all the ideas that partake of their associative whole. Freud refers to them by the term of purposive ideas (*représentations-but*): cf. *The Interpretation of Dreams*, Ch. VII, p. 512.
6 By vicissitudes (*Triebschicksale*), Freud means the different fates or destinies that the drive imposes on itself to make up for its inherent failure to achieve its aim, that is to say its failed satisfaction. There are five such vicissitudes: repression, sublimation, turning around into its opposite, turning round upon the subject's own self, and finally the passage from activity to passivity (cf. Freud, S., *Papers on Metapsychology*, 1915a). The reader will have noticed in the course of reading the present book that as far as the oneiric drive is concerned the above-mentioned list is not exhaustive.
7 See above, Part One, Chapter Three, "*L'objet 'petit-âme'*, what is closest and yet most distant".

INDEX OF DREAMS

BIBLIOGRAPHY

Adler, A. (1925). *The Practice and Theory of Individual Psychology*. P. Radin (Trans.). London: Routledge & Kegan Paul.

Agamben, G. (1997). *Homo sacer, le pouvoir souverain et la vie nue*. Paris: Seuil.

Akert, K., Bally, C. and Schadé, J. (1965). (Eds). *Sleep Mechanisms*. Amsterdam: Elsevier Science Ltd.

Antrobus, J., Ehrlichman, H. and Weiner, M. (1978). EEG asymmetry during REM and NREM: Failure to replicate. *Sleep Research*, 7(24): 359–368.

Aristotle (1936). *Physics*. London: Clarendon Press.

Aristotle (1952). *De Anima* (On the Soul). J. A. Smith (Trans.). Chicago, IL: Encyclopaedia Britannica.

Aserinsky, E. and Kleitman, N. (1953). Regularly occurring periods of eye motility, and concurrent phenomena, during sleep. *Science*, 118: 273–274.

Aserinsky, E. (1965). Brain wave pattern during the rapid eye movement period of sleep. *The Physiologist*, 8: 25–97.

Atkinson, Q. D. (2011). Phonemic diversity supports a serial founder effect model of language expansion from Africa. *Science*, 332: 346–349.

Bateson, P. P. G. (1966). The characteristics and context of imprinting. *Biological Reviews*, 41: 177–220.

Bennett, D. (2010). *Content and Consciousness*. London: Routledge.

Bennett, M. and Hacker, P. (2003). *Philosophical Foundations of Neuroscience*. Oxford: Blackwell.

Bennett, M., Bennett, D., Hacker, P. and Searle, J. (2007). *Neuroscience and Philosophy, Brain, Mind, and Language*. Columbia, NY: Columbia University Press.

Bensch, C. (2000). *Jeux de velus*. Paris: Odile Jacob.

Bergson, H. (1965). *Duration and simultaneity*. Indianapolis: Bobbs-Merrill.

Bernard, C. (1957). (1865). *Introduction to the Study of Experimental Medicine*. H. Copley Green (Trans.). New York, NY: Dover.

Bernstein, N. A. (1966). *On Dexterity and Its Development*. K. L. Latash and M. T. Turvey (Eds). Mahwah, NJ: Erlbaum Associates.

Bernstein, N. A. (1967). The problem of the interrelationships between coordination and localization. In: Bernstein, N. A. (Ed.), *The Coordination and Regulation of Movements*. Oxford: Pergamon Press, pp. 15–59.

Berthoz, A. (2000). *The Brain's Sense of Movement*. Cambridge, MA: Harvard University Press.

Berthoz, A. (2013). *La décision*. Paris: Odile Jacob.

Berthoz, A. and Petit, J.-L. (2006). [2008]. *The Physiology and Phenomenology of Action*. C. McCann (Trans.). Oxford: Oxford University Press.

Besson, M. and Schön, D. (2001). Comparison between language and music. In: R. Zatorre and I. Peretz (Eds). *The Biological Foundations of Music*, 930: 232–258. New York, NY: Annals of the New York Academy of Sciences.

Bichat, X. and Pichot, A. (1994). *Recherches physiologiques sur la vie et la mort*. Paris: Flammarion.

Bloch, O. and von Wartburg, W. (2008). *Dictionnaire étymologique de la langue française*. Paris: Presses Universitaires de France.

Boss, M. (1957). *Analysis of Dreams*. London: Rider & Co.

Bourguignon, A. (1998). *Psychopathologie et épistémologie*. Paris: Presses Universitaires de France.

Bowlby, J. (1969). *Attachment: Attachment and Loss (Vol 1)*. New York, NY: Basic Books.

Bowlby, J. (1973). *Separation: Anxiety and Anger. Attachment and Loss (Vol 2)*. London: Hogarth.

Bowlby, J. (1980). *Loss: Sadness and Depression. Attachment and Loss (Vol 3)*. London: Hogarth.

Broughton, R. (1975). Biorhythmic variations in consciousness and psychological function. *Canadian Psychological Review*, 16: 217–239.

Brown, G. (1912). *Rhythmic Movements: A Contribution to the Study of the Central Nervous System*. Edinburgh: Thesis, University of Edinburgh.

Buytendijk, F. (1920). [1928]. *Psychologie der dieren [Animal Psychology] I and II*. Haarlem: Eerven, F., Bohn.

Caillois, R. (1956). *L'incertitude qui vient des rêves*. Paris: Gallimard.

Caillois, R. (1992). *Les jeux et les hommes*. Paris: Gallimard.

Campos-Bueno, J. and Martin-Araguz, A. (2012). Neuron Doctrine and Conditional Reflexes at the XIV International Medical Congress of Madrid of 1903. *Psychologia Latina*, 3(1): 10–22.

Canguilhem, G. (1955). *La formation du concept de réflexe aux XVIIe et XVIIIe siècles*. Paris: Presses Universitaires de France.

Canguilhem, G. (2002). *Écrits sur la médecine*. Paris: Seuil.

Cartwright, R. D. (1999). Problem solving: waking and dreaming. *Journal of Abnormal Psychology*, 83(4): 451–455.

Cate, C. and Slater, P. J. B. (1991). Song learning in zebra finches: how are elements from two tutors integrated? *Animal Behaviour*, 42: 150–152.

Changeux, J.-P. (1983). *L'homme neuronal*. Paris: Arthème Fayard.

Changeux, J.-P. and Ricœur, P. (2000). *Ce qui nous fait penser, la nature et la règle*. Paris: Odile Jacob.

Charles, M. and Lebrun, J.-P. (2009). *La nouvelle économie psychique: la façon de penser et de jouir aujourd'hui*. Paris: Erès.

Chemama, R. and Vandermersch, B. (Eds). (1998). *Dictionnaire de la psychanalyse*. Paris: Larousse.

Chomsky, N. (1975). *Reflections on Language*. New York, NY: Pantheon Books.

Chomsky, N. (2011). *Le langage et la pensée*. Paris: Payot.

Cohen, A. and Varela F. (2000). Facing up to the embarrassment: the practice of subjectivity in neuroscientific and psychoanalytic experience. *Journal of European Psychoanalysis*, 10–11: 41–55.

Comte, A. (1996). *Philosophie des sciences*. Paris: Gallimard.

Corbin, H. (1971). *En Islam iranien*. Paris: Gallimard (Collection Tel), vol. I.

Corsi, P. (2001). *Lamarck, genèse et enjeux du transformisme, 1710–1830*. Paris: CNRS.

Craig, W. (1918). Appetites and aversions as constituents of instincts. *Biological Bulletin*, 34(2): 91–107.

Crick, F. and Mitchinson, G. (1995). REM sleep and neural nets. *Behavioral Brain Research*, 69: 147–155.

Cuvelier, A. (1992). (Ed.). *Psychisme et intelligence artificielle*. Nancy: Presses Universitaires de Nancy.

Cynx, J. (1990). Experimental determination of a unit of song production in the zebra finch (Taeniopygia guttata). *Journal of Comparative Psychology*, 104: 3–10.

Damasio, A. (1994). *Descartes' Error: Emotion, Reason, and the Human Brain*. London: Macmillan.

Damasio, A. (2000). *The Feeling of What Happens: Body and Emotion in the Making of Consciousness*. New York, NY: Harcourt Inc.

Damasio, A. (2003). *Looking for Spinoza: Joy, Sorrow, and the Feeling Brain*. New York, NY: Harcourt Inc.

Demazeux, S. (2013). *Qu'est-ce que le DSM? Genèse et transformations de la Bible américain de la psychiatre*. Paris: Editions d'Ithaque.

Dement, W. C. (1958). The occurrence of low voltage, fast, electroencephalogram patterns during behavioral sleep in the cat. *Electroencephalography and Clinical Neurophysiology*, 10(2): 291–296.

Dement, W. C. (1965). An essay on dreams: the role of physiology in understanding their nature. In: *New Directions in Psychology II*. New York, NY: Holt, Rinehart & Winston Inc., pp. 137–257.

Dement, W. C. (1972). *Some Must Watch While Some Must Sleep*. New York, NY: Scribner.

Dennett, D. (2007). Philosophy as native anthropology. In: Daniel Robinson (Ed). *Neuroscience and Philosophy, Brain, Mind and Language*. Columbia, NY: Columbia University Press.

Depraz, N., Varela, F. and Vermersch, P. (2003). *On Becoming Aware: A Pragmatics of Experiencing*. Amsterdam: J. Benjamins.

Derrida, J. (1967). [1978]. *Writing and Difference*. Alan Bass (Trans.). London: Routledge & Kegan Paul.

Descartes, R. (1662). [1972]. *Treatise of Man*. T. S. Hall (Trans.). Cambridge, MA: Harvard University.

Descombes, V. (1995). *La denrée mentale*. Paris: Les Editions de Minuit.

Descombes, V. (2011). *The Mind's Provisions: A Critique of Cognitivism*. Princeton, NJ: Princeton University Press.

Dolto, F. (1984). *L'image inconscient du corps*. Paris: Seuil.

Doupe, A. and Kuhl, P. K. (1999). Birdsong and human speech: common themes and mechanisms. *Annual Review of Neuroscience*, 22: 567–631.

Dreyfus, H. (Ed.). (1982). *Husserl, Intentionality and Cognitive Science*. Cambridge, MA: MIT Press.

Ducrot, O. and Todorov, T. (1972). *Dictionnaire encyclopédique des sciences du langage*. Paris: Editions du Seuil (Points). [(1979). *The Encyclopaedic Dictionary of the Sciences of Language*. Catherine Porter (Trans.). Baltimore, MD: Johns Hopkins University Press.]

Dupont, J.-C. (2011). *History of Neuroscience in France and Russia*. Paris: Editions Hermann.

Erderlyi, M. H. (1985). *Psychoanalysis: Freud's Cognitive Psychology*. New York, NY: W. H. Freeman.

Feuerhahn, W. (2011). Un tournant neurocognitiviste en phénoménologie? Sur l'acclimatation des neurosciences dans le paysage philosophique français. *Revue d'Histoire des Sciences Humaines*, 25: 59–79.

Flanagan, O. (1995). Deconstructing dreams, the spandrels of sleep. *The Journal of Philosophy*, 92(1): 5–27.

Fodor, J. (1980). *The Language of Thought*. Cambridge, MA: Harvard University Press.

Foucault, M. (1966). *Les mots et les choses*. Paris: Gallimard.

Foucault, M. (2004). *The Birth of Biopolitics: Lectures at the Collège de France, 1978–1979*. London: St Martin's Press.

Foulkes, D. and Fleischer, S. (1975). Mental activity in relaxed wakefulness. *Journal of Abnormal Psychology*, 84(1): 66–75.

Foulkes, D. (1993). Dreaming and REM sleep. *Journal of Sleep Research*, 2: 199–202.

Frege, G. (1971). *Ecrits logiques et philosophiques*. Paris: Seuil.

Freud, S. (1891). [1953]. *On Aphasia*. E. Stengel (Trans.). New York, NY: International Universities Press.

Freud, S. (1900). *The Interpretation of Dreams. S.E.*, 4–5. London: Hogarth, pp. 1–600.

Freud, S. (1901). On dreams. In: *The Interpretation of Dreams. S.E.*, 5. London: Hogarth, pp. 339–686.

Freud, S. (1905). Three Essays on Sexuality. *S.E.*, 7. London: Hogarth, pp. 135–243.

Freud, S. (1910). [1911]. Psycho-analytic notes on an autobiographical account of a case of paranoia (dementia paranoides). *S.E.*, 12. London: Hogarth, pp. 1–82.

Freud, S. (1912). On the universal tendency to debasement in the sphere of love. *S.E.*, 11. London: Hogarth, pp. 177–190.

Freud, S. (1912–1913). Totem and Taboo. *S.E.*, 13. London: Hogarth, pp. 1–161.

Freud, S. (1914). [1918]. From the history of an infantile neurosis. *S.E.*, 17. London: Hogarth, pp. 1–122.

Freud, S. (1915a). Papers on Metapsychology. *S.E.*, 14. London: Hogarth.

Freud, S. (1915b). Instincts and their Vicissitudes. *S.E.*, 14. London: Hogarth, pp. 117–140.

Freud, S. (1920). Beyond the Pleasure Principle. *S.E.*, 18. London: Hogarth, pp. 1–64.

Freud, S. (1923). The Ego and the Id. *S.E.*, 19. London: Hogarth, pp. 3–66.

Freud, S. (1925a). [1924]. A note upon the mystic writing-pad. *S.E.*, 19. London: Hogarth, pp. 227–232.

Freud, S. (1925b). Inhibition, Symptoms and Anxiety. *S.E.*, 20. London: Hogarth, pp. 75–174.

Freud, S. (1925c). Negation. *S.E.*, 19. London: Hogarth, pp. 233–239.

Freud, S. (1930). Civilization and Its Discontents. *S.E.*, 21. London: Hogarth, pp. 59–145.

Freud, S. (1950). [1895]. Project for a Scientific Psychology. *S.E.*, 1. London: Hogarth: pp. 281–397.

Freud, S. (1982). *Die Traumdeutung*. Studienausgabe Band II, Fischer Wissenschaft.

Freud, S. (1985). *The Complete Letters of Sigmund Freud to Wilhelm Fliess, 1887–1904*. J. M. Masson (Ed.). London: The Belknap Press and Cambridge, MA: Harvard University Press.

Genette, G. (1972*). Figures III*. Paris: Le Seuil,

Gide, A. (1939). *Journal 1889–1939*. Paris: Gallimard, Pléiade, vol 1.

Gluck, M. A., Anderson, J. R. and Kosslyn, S. M. (2007). *Memory and Mind: A Festschrift for Gordon H. Bower*. London: Psychology Press.

Goldstein, L., Stolzfus, N. and Gardocki, J. (1972). Changes in interhemispheric amplitude relations in EEG during sleep. *Physiological Behavior*, 8: 811–815.

Goldstein, K. (1948). *Language and Language Disturbances: Aphasic Symptom Complexes and Their Significance for Medicine and Theory of Language*. New York, NY: Grune & Stratton.

Gori, R. and Del Volgo, M-J. (2005). *La santé totalitaire, essai sur la médicalisation de l'existence*. Paris: Flammarion (Champ essais).

Gori, R. (2011). *La dignité de penser, les liens qui libèrent*. Paris: Actes Sud.

Harrison, J. (2002). *Off to the Side: A Memoir*. New York, NY: Grove Press.

Hartmann, E. (1998). *Dreams and Nightmares: The New Theory on the Origin of Dreams*. New York, NY: Plenum.

Hauser, M. D., Chomsky, N. and Fitch, W. (2002). The faculty of language: what is it, who has it, and how did it evolve? *Science*, 298: 1569–1579.

Head, H. (1926). *Aphasias and Kindred Disorders of Speech*. Cambridge: Cambridge University Press.

Heidegger, M. (1927). [1962]. *Being and Time*. London: Blackwell.

Heidegger, M. (1982). *The Basic Problems of Phenomenology*. Albert Hofstadter (Trans.). Bloomington and Indianapolis: Indiana University Press.

Herrnstein, R. J. and Murray, C. (1996). *The Bell Curve: Intelligence and Class Structure in American Life*: New York, NY: Free Press.

Hinde, Robert A. (1963). The nature of imprinting. In: B. M. Boss (Ed.). *Determinants of Infant Behaviour, vol. 2*. London: Methuen, pp. 227–233.

Hirshkowitz, M., Turner, D., Ware, J. and Karacan, I. (1979). Integrated EEG amplitude asymmetry during sleep. *Sleep Research*, 8: 25.

Hobson, J. A. (1988). *The Dreaming Brain*. New York, NY: Basic Books.

Hobson, J. A. (2002). *Dreaming: A Very Short Introduction*. Oxford: Oxford University Press.

Huizinga, J. (1938). [1949]. *Homo Ludens: A Study of the Play-Element in Culture*. London: Routledge & Kegan Paul.

Hume, D. (1738). [2014]. *A Treatise on Human Nature: Being an Attempt to Introduce the Experimental Method of Reasoning into Moral Subjects*. London: John Noon.

Huneman, P. (1998). *Bichat, la vie et la mort*. Paris: Presses Universitaires de France.

Husserl, E. (1964). *Leçons pour une phénoménologie de la conscience intime du temps*. Henri Dussort (Trans.). Paris: Presses Universitaires de France.

INSERM (2004) (Eds). *Psychothérapie: Trois approches évaluées. Rapport (Expertise collective)*. Paris: Inserm. Available at: http://www.ipubli.inserm.fr/handle/10608/57.

Jackson, F. (1982). Epiphenomenal Qualia. *The Philosophical Quarterly*, 32(127): 127–136.

Jasper, H., Ricci, G. F. and Doane, B. (1958). Patterns of cortical neuron discharge during conditioned responses in monkeys. In: G. E. W. Wolstenholme and C. M. O'Connor (Eds). *Neurological Basis of Behavior*. London: J. & A. Churchill Ltd.

Jones, E. (1957). [1980]. *Sigmund Freud, Life and Work, vol. I*. London: Hogarth.

Jones, J. G. and Stuart, D. G. (2011). Thomas Graham Brown (1882–1965): Behind the scenes at the Cardiff Institute of Physiology. *Journal of the History of the Neurosciences*, 20(3): 188–209.

Jouvet, Michel. (1993). [2001]. *The Paradox of Sleep: The Story of Dreaming*. L. Garey (Trans.). Cambridge, MA: MIT Press.

Jung, C. (2010). *Children's Dreams: Notes from the Seminar Given in 1936–1940*. Maria Meyer-Grass and Lorenz Jung (Eds). Ernst Falzeder with the collaboration of Tony Woolfson (Trans.). Princeton, NJ: Princeton University Press.

Kant, E. (1951). [1791]. *The Critique of Judgement*. J. H Bernard (Trans.). New York, NY: Hafner Publishing.

Kaplan-Solms, K. and Solms, M. (2000). *Clinical Studies in Neuro-Psychoanalysis, Introduction to a Depth Neuropsychology*. London: Karnac Books.

Kenagy, J. (2009). *Designed to Adapt: Leading Healthcare in Challenging Times*. Bozeman, MT: Second River Healthcare Press.

Klein, M. (1957). *Envy and Gratitude and Other Papers, 1921–45*. In: The Writings of Melanie Klein, Vol. III. London: Hogarth.

Kleitman, N. (1963). *Sleep and Wakefulness*. Chicago, IL: University of Chicago Press.

Kosslyn, S. M. and Koering, O. (1995). *Wet Mind*. New York, NY: The Free Press.

Kozulin, A. (1984). *Psychology in Utopia: Toward a Social History of Soviet Psychology*. Cambridge, MA: MIT Press.

Kramer, M. (1993). The selective mood regulatory function of dreaming. In: A. Moffitt, M. Kramer and R. Hoffmann (Eds). *The Functions of Dreaming*. Albany, NY: State University of New York, pp. 139–195.

Lacan, J. (1960–1961). *Le transfert*. (Seminar of 19 April, 1961.) Paris: Seuil.

Lacan, J. (1961–1962). *Identification. Seminar IV*. Unpublished.

Lacan, J. (1966a). [2006]. Science and truth. In: *Écrits*, B. Fink (Trans.). New York, NY: Norton.

Lacan, J. (1966b). [2006]. The Purloined Letter. In: *Écrits*. B. Fink (Trans.). New York, NY: Norton.

Lacan, J. (1966c). [2006]. The subversion of the subject and the dialectic of desire. *Écrits*. B. Fink (Trans.). New York, NY: Norton.

Lacan, J. (1967). *The Seminar, Book XIV, The Logic of Fantasy, lesson of 12 avril 1967*. Unpublished.

Lacan, J. (1973a). [1977]. *The Four Fundamental Concepts of Psycho-Analysis*. London: Hogarth.

Lacan, J. (1973b) L´étourdit. *Scilicet* 4: 5–52. Paris: Seuil.

Lacan, J. (1974). [1990]. *Television*. D. Hollier, R. Kraus, and A. Michelson (Trans.). New York, NY: W. W. Norton & Co.

Lacan, J. (1975a). *Encore 1972–1973: The Seminar of Jacques Lacan, Book XX*. J.-A. Miller (Ed.). B. Fink (Trans.). New York, NY: Norton.

Lacan, J. (1975b). [1988]. *Freud's Papers on Technique. Seminar I*. J.-A. Miller (Ed.). J. Forrester (Trans.). New York, NY: Norton.

Lacan, J. (1977). *The Ego in Freud's Theory and in the Technique of Psychoanalysis, 1954–55*. S. Thomaselli (Trans.). Cambridge: Cambridge University Press, 1988.

Lamarck, J.-B. (1964). [1802]. *Hydreology*. A. V. Carozzi (Trans.). University of Illinois, Urbana.

Lambert, X. (Ed.). (2011). *Le post-humain et les enjeux du sujet*. Paris: Harmattan.

Laplanche, J. and Pontalis, J-B. (1967). [1973]. *The Language of Psychoanalysis*. D. Nicholson-Smith (Trans.). London: Hogarth.

Lashley, K. (1929). *Brain Mechanisms and Intelligence*. Chicago, IL: University of Chicago Press.

Lemerle, S. (2014). *Le singe, le gène et le neurone*. Paris: Presses Universitaires de France.

Lévi-Strauss, C. (1949). [1969]. *The Elementary Structures of Kinship*. Needham, R (Ed.). J. Bell, J. R. von Sturmer (Trans.). Boston, MA: Beacon Press.

Libera, A. de. (2007). *Archéologie du sujet, naissance du sujet*. Paris: Vrin.

Llinás, R. (2002). *I of the Vortex*. Cambridge, MA: MIT Press.

Llinás, R. R. and Paré, D. (1991). Commentary of dreaming and wakefulness. *Neuroscience*, 44(3): 521–535.

Llinás, Rodolfo. (1988). The intrinsic electrophysiological properties of mammalian neurons: insight into central nervous system function. *Science*, 242: 1654–1664.

Llinás, R. and Ribary, U. (1993). Coherent 40-Hz. oscillation characterizes dream state in humans. *Proceedings of the National Academy of Sciences*, 90: 2078–2081.

Locke, J. (1689). [1824]. An essay concerning human understanding. In: *The Works of John Locke in Nine Volumes*, Vols. 1 & 2: London: Rivington.

Locke, J. (1690). [1824]. Two treatises on civil government. In: *The Works of John Locke in Nine Volumes*, Vol. 4: London: Rivington.

Lorenz, K. (1935). Der Kumpan in der Umwelt des Vogels. *Journal für Ornithologie*, 83: 137–213.

Lorenz, K. (1956). Plays and vacuum activities. In: M. Autuori (Ed.). *L'instinct dans le comportement des animaux et de l'homme*. Paris: Masson, pp. 633–645.

Lorenz, K. (1970*). Studies in Animal and Human Behaviour*. Cambridge, MA: Harvard University Press.

Lorenz, K. (1978). [1981]. *The Foundations of Ethology*. K. Z. Lorenz and R. Warren Kickert (Trans.). New York, NY: Springer Verlag.

Luria, A. R. (1987). Mind and brain: Luria's philosophy. In: R. L. Gregory (Ed.). *The Oxford Companion to the Mind*. Oxford & New York: Oxford University Press.

Maijer, O. G. and Bruijn, S. M. (2007). The loyal dissident: Bernstein and the double-edged sword of Stalinism. *Journal of the History of the Neurosciences*, 16: 206–224.

Marey, E. J. (1895). [April 2011]. The physiological station at Paris. In: *Congressional Edition*, vol. 3341, U.S. Government Printing Office, digital version,.

Marr, D., Ullman, S. and Paggio, T. (2010). *Vision*. Cambridge, MA: MIT Press.

Marty, P and M'Uzan, M. de (1963). La pensée opératoire. *Revue française de psychanalyse*, 27 (special edition): 345–366. Congress for French-speaking analysts, Barcelona, 8–11 June, 1962.

Marty, P. (1998). *Les mouvements individuels de vie et de mort*. Paris: Payot.

Mazauric, S. (1998). *Gassendi, Pascal et la querelle du vide*. Paris: Presses Universitaires de France.

McCarley, R. (1998). Dreams: disguise of forbidden wishes or transparent reflections on a distinct brain state. *Annals of the New York Academy of Sciences*, 843: 116–133.

McCarley, R. W. and Hobson, J. A. (1977). The neurological origins of psychoanalytic dream theory. *American Journal of Psychiatry*, 13(11): 1211–1221.

Meijer, O. G. and Bongaardt, R. (1998). Bernstein's last paper: the immediate tasks of neurophysiology in the light of the modern theory of biological activity. *Motor Control*, 2: 2–9.

Melman, C. and Lebrun, J.-P. (2009). *La nouvelle économie psychique: la façon de penser et de jouir aujourd'hui*. Paris: Erès.

Merleau-Ponty, M. (1945). [2002]. *Phenomenology of Perception*. London: Routledge.

Merleau-Ponty, M. (1964). [1968]. *The Visible and the Invisible*. Illinois: Northwestern University Press.

Montangero, J. (1993). Dream, problem-solving and creativity. In: C. Cavallero and D. Foulkes (Eds). *Dreaming as cognition*. New York, NY: Harvester Wheatsheaf, pp. 93–113.

Movallali, K. (1988). Questionner la dénégation. *Littoral*, 25: 129–140.

Muller, M. (1965). [1999]. *Les voix narratives dans A la Recherche du Temps Perdu*. Geneva: Droz.

Nagel, T. (1974). What Is It Like to Be a Bat? *The Philosophical Review*, 83(4): 435–450.

Neidich, Warren. (Ed.). (2013). *The Psychopathologies of Cognitive Capitalism: Part Two*. Berlin: Archive Books.

Nielsen, Tore A. (2003). A review of mentation in REM and NREM sleep: "cover" REM sleep as a possible reconciliation of two opposing models. In: Edward F. Pace-Schott, Mark Solms, Mark Blagrove and Stevan Harnad (Eds). *Sleep and Dreaming.* Cambridge: Cambridge University Press, pp. 59–74.

Neisser, U. (1966). *Cognitive Psychology.* NJ: Prentice Hall.

Neisser, U. (1976). *Cognition and Reality: Principles and Implications of Cognitive Psychology.* New York, NY: W. H. Freeman & Co. Ltd.

Norton, D. F. and Taylor, J. (Eds). (2008). *The Cambridge Companion to Hume.* Cambridge: Cambridge University Press.

Ouelbani, Malika. (2006). *Le Cercle de Vienne.* Paris: Presses Universitaires de France.

Palmer, J. D. (1988). Clockers. *The Sciences,* 38(5): 30–34.

Petitot, J. (1983). *Sémiotique et théorie des catastrophes.* Paris: Centre National de la Recherche Scientifique.

Petitot, J. (1993). Phénoménologie naturalisée et morphodynamique: la fonction cognitive du synthétique *a priori. Intellectica,* 17: 79–126.

Petitot, J. (2003). *Morphologie et esthétique.* Paris: Maisonneuve et Larose.

Petitot, J. (2009). *Neurogéométrie de la vision: modèles mathématiques et physiques des architectures fonctionnelles.* Paris: Editions Ecole Polytechnique.

Petitot, J. and Thom, R. (1996). *Sémiotique et théorie des catastrophes.* Limoges: Presses Universitaires de Limoges.

Piccin, Alberto, Couchman, M., Clayton, J., Chalmers, D., Costa, R. and Kyriacou, C. (2000). The clock gene period of the housefly, Musca domestica, rescues behavioral rythmicity in Drosophila melanogaster: Evidence for intermolecular co-evolution? *Genetics,* 154: 747–758.

Pickenhein, L. (1980). A neuroscientist's view on theories of complex movement behaviour. In: O. G. Meijer and K. Roth (Eds). *Complex Movement Behaviour: The motor-action controversy.* Amsterdam: North-Holland, pp. 463–487.

Pommier, G. (2010). *Comment les neurosciences* démontrent la psychanalyse. Paris: Flammarion.

Post, R. (1992). Transduction of psychosocial stress into the neurobiology of recurrent affective disorder. *Am J. Psychiatry,* 149(8): 999–1010.

Procháska, G. (1851). Dissertation on the functions of the nervous system. In: J. A. Unzer (Ed.). *The Principles of Physiology.* London: Sydenham Society.

Proust, M. (1913–1927). *La Recherche du Temps Perdu.* Paris: Gallimard.

Renck, J.-L. and Servais, V. (2002). *L'éthologie.* Paris: Editions du Seuil (Points).

Revonsuo, A. (2000). Reinterpretation of dreams: an evolutionary hypothesis of the function of dreaming. *Behavioral and Brain Sciences,* 23: 793–1121.

Rey, A. (Ed.). (1992). *Dictionnaire historique de la langue française.* Paris: Editions du Robert.

Roland, G. (2011). *La dignité de penser, les liens qui libèrent.* Paris: Denoël.

Russell, B. (1997). [1912]. *The Problems of Philosophy.* Oxford: Oxford University Press.

Saint Augustine (Bishop of Hippo). (1960). *The Confessions of St Augustine.* New York, NY: Image.

Sami-Ali, M. (1980). *Le banal.* Paris: Gallimard.

Sartre, J.-P. (1940). [2004]. *The Imaginary: A Phenomenological Psychology of the Imagination.* J. Webber (Trans.). London: Routledge.

Sartre, J.-P. (1943). *Being and Nothingness.* H. Barnes (Ed.). New York, NY: Washington Square.

Searle, J. (1980). Minds, Brains, and Programs. *Behavioral and Brain Sciences,* 3: 417–424.

Searle, J. (1992). *The Rediscovery of the Mind.* Cambridge, MA: MIT Press.

Sechenov, I. (1957). *Œuvres philosophiques et psychologiques choisies*. Moscow: Éditions en Langues Étrangères.

Sherrington, C. S. (1942). [1951]. *Man on his Nature*. Cambridge: Cambridge University Press.

Singer, W. (1977). Extraretinal influences to the thalamus. *Physiological Review*, 57: 386–420.

Singer, W. and Varela, F. J. (1987). Neuronal dynamics in the visual corticothalamic pathway revealed through binocular rivalry. *Experimental Brain Research*, 66: 10–20.

Sluckin, W. (1965). *Imprinting and Early Learning*. London: Methuen.

Solms, M. (1997). *The Neuropsychology of Dreams*. Mahwah, NJ: Lawrence Erlbaum Associates,

Solms, M. (2000). Dreaming and REM sleep are controlled by different brain mechanisms. *Behavioral and Brain Sciences*, 23(6): 843–50.

Spitz, R. A. (1945). Hospitalism. *The Psychoanalytic Study of the Child*, 2: 113–117.

Sprague, R. (1980). A framework for the development of decision support systems. *Management Information Systems (MIS) Quarterly*, 4(4): 1–25.

Steiner, G. (1989). *Real Presences*. Chicago, IL: University of Chicago Press.

Sugerman, A., Goldstein, L., Marjerrison, G. and Stolzfus, N. (1973). Recent research in EEG amplitude analysis. *Disorders of the Nervous System*, 34: 162–166.

Sulloway, F. (1979). *Freud, Biologist of the Mind: Beyond the Psychoanalytic Legend*. New York, NY: Basic Books.

Sztulman, H., Barbier, A. and Caïn, J. (Eds). (1986). *Les fantasmes originaires*. Paris: Éditions Privat.

Thèves, P. and This, B. (Trans.). (1982). *La dénégation*. Paris: Le Coq Héron.

Tinbergen, N. (1966). *Le comportement animal*. Paris: Collections Life.

Tore, A. N. (2003). A review of mentation in REM and NREM sleep: "cover" REM sleep as a possible reconciliation of two opposing models. In: Edward F. Pace-Schott, Mark Solms, Mark Blagrove and Stevan Harnad (Eds). *Sleep and Dreaming*. Cambridge: Cambridge University Press.

Tye, M. (1991). *The Imagery Debate*. Cambridge, MA: MIT Press.

Valéry, P. (1932). *Choses tues*. Paris: Gallimard.

Varela, F. (1996a). Neurophenomenology: a methodological remedy for the hard problem. *Journal of Consciousness Studies*, 3: 330–350.

Varela, F. (1996b). *Invitation aux sciences cognitives*. Paris: Edition du Seuil (Points-Sciences).

Varela, F., Thompson, E. and Rosche, E. (1991). *The embodied mind: cognitive science and human experience*. V. Havelange (Trans.). Cambridge, MA: MIT Press.

Velluti, R. (2010). *The Auditory System in Sleep*. London: CITYAcademic Press.

Vogel, G. (1978). An alternative view of the neurobiology of dreaming. *American Journal of Psychiatry*, 135(3): 1531–5.

Wilson, E. O. (1975). *Sociobiology: The New Synthesis*. Cambridge, MA: Harvard University Press.

Winnicott, D. W. (1953). Transitional objects and transitional phenomena. *International Journal of Psychoanalysis*, 34(2): 89–97.

Winnicott, D. W. (1971). *Playing and Reality*. London: Tavistock.

GENERAL INDEX

Milton Keynes UK
Ingram Content Group UK Ltd.
UKHW030902141024
449569UK00026B/1317

9 781138 858251